LEAVING WAR

LEAVING WAR

FROM AFGHANISTAN'S PECH VALLEY TO HADRIAN'S WALL:

A VETERAN'S SEARCH FOR PEACE

ANTONIO SALINAS

DEEDS PUBLISHING | ATHENS

Published by Deeds Publishing in Athens, GA
www.deedspublishing.com

Printed in The United States of America

Cover and interior design by Deeds Publishing

ISBN 978-1-961505-47-6

Books are available in quantity for promotional or premium use.
For information, email info@deedspublishing.com.

First Edition, 2025

10 9 8 7 6 5 4 3 2 1

For those who carry the war inside them still, who continue to search for peace. You are not alone.

For those who have walked Hadrian's Wall, or dream of doing so — come to the wall not only to hear the heartbeat of the Roman Empire ... but to listen for your own heart and the quiet stirrings of your dreams.

To my family — Claire, Roman, Shannon, and Conan.

CONTENTS

SHADOWS OF WAR. HUDSON VALLEY HIGHLANDS. WINTER 2017.

Sometimes, I need to be alone in the wilderness. I wake up in the pre-dawn twilight and stare at my phone, which shows 5:30 am — too early for most, but the perfect time for those who have hunted enemy soldiers. I calmly dress in the darkness and enjoy a warm cup of coffee, hoping the heat and caffeine will help me to ignore old wounds and the arthritis that slowly crept into my joints. I sling my backpack over my shoulder, step into the cold embrace of solitude, and head for my truck.

The mountains call to me. They've always been sanctuaries for warriors seeking something they cannot name. I drove a short distance to the nearby mountains and picked a trail where I knew I wouldn't see another soul for hours. I only want to hear the chirps of birds, the creaking of trees, and the crunching of light snow under my boots. I want peace, at least for a few hours.

I find a secluded trail with only a few, well-frozen footprints. I begin my hike just as rosy-fingered dawn graced the sky. The dawn is sacred to me, as it is to all those who have practiced the sanguine profession of war. This was a period when numerous battles began in Afghanistan. Our guerrilla opponents waited for the dawn's light to start their attacks against us. Yet, today, there would be no horrid screams of bullets whizzing overhead. This morning, it was just me.

I stare at my breath rising in the cold winter air as if climbing on some invisible stairwell. It was like small parts of my soul leaving my body and returning, like a prayer to the heavens, floating. I watch the waves of my organic-made smoke float weightlessly higher and higher. My backpack was heavy, but this time not with the provisions of war. Instead, it was filled with extra boots, socks, water, and even a few dumbbells to give it more weight. I remember when that extra weight was made up of additional 7.62 ammunition, some magazines filled with 5.56 ammo, batteries, a VS-17 panel (a small bright orange flag), and a couple of Meals Ready to Eat (MREs).

As I climb higher, my sweat mixed with the icy air, and the burn in my muscles brought me back to the battlefield. My breath, visible in the cold, rises like a prayer to the heavens. For a moment, I was no longer on this mountain trail in the Hudson Valley Highlands. Instead, I was back there. I was back in the Pech Valley of Kunar Province, Afghanistan.

I am back in my valley... and I am no longer alone.

I am back on patrol.

I hear my men as they walked beside me. I listen to the subtle noise of boots stepping on rocks and the low creaking of their heavy Kevlar armor. I hear them begin to breathe a bit quicker, watching their lungs rise and fall as mine did. Intermingled in and amongst the trees, I see their faces. I do not see them as bearded veterans, but rather as they were. I see them as they were then, like powerfully muscled, armored wolves searching for prey. I make eye contact with them as they turn their heads to check their "6" on patrol. I wish that I could go back there and join them, back before drugs or suicide stole some of them away. I wish I could save them here in this land of peace as I did at war.

Soon, they fade away. The sand and Afghan dirt turned to snow again.

At the mountain's summit, I paused for a few moments, my chest heaving from exertion, and I am alone with only my thoughts. The highlands open before me, breathtaking and infinite, a world far removed from the battlefield. For the first time in what felt like an eternity, I felt peace creeping into the corners of my exhausted soul. Yet, it didn't stay. The mountains reminded me of what I am — a warrior forever scarred by war, carrying its echoes even into the stillness of the wild.

A few hours later, I finished my hike and threw my heavy backpack into the back of my truck. I drove home and tried not to be sucked in by the seductive sights of the snow-covered highlands. I pull into my driveway, cut the engine, and wait a few moments, enjoying the serenity that only silence and solitude offer a warrior. I had to go home. I entered my house and kissed my wife and toddler.

"Did you hike very far, Daddy?" He asked.

"I went a bit further than you would ever guess." I smile in reply.

I miss my soldiers, and I miss my war.

You are not alone.

CHAPTER ONE

WALKING AWAY FROM WAR

It has been nearly fifteen years since I returned from my first combat deployment in the Pech River Valley, Kunar Province, Afghanistan, from 2009 to 2010. I spent most of that year as an infantry officer, guiding young Americans through the unforgiving crucible of my generation's war. Those months forged memories I can never forget, along with physical and emotional scars I will always carry.

I survived well over a hundred firefights, narrowly escaping death more times than I can count—dodging bullets, evading RPGs, mortars, 107mm rockets, and even limping away from an IED (improvised explosive device) blast. Yet, it's not just the toll of near-death experiences that lingers. It's also the very tough choices that I had to make. To survive and protect the lives of Soldiers under my charge, I ordered the slaughter of my fellow man. In those god-forsaken moments, I met a part of myself I never thought existed. I will forever remember hunting Man—the primal, unrelenting force within me that emerged in the chaos of war.

And yet, even after all this time, I still try to walk away from it. You can leave war, but I wonder if it ever truly leaves us.

Hiking and walking serve as my perpetual path for my soul to find peace. These steps always feel like a silent rebellion against the ghosts of war, an unspoken hope that distance might somehow loosen their grip on my soul. In June of 2010, I thought I could simply walk it off—strip the weight of combat from my shoulders by trekking the trail next to the ancient stones of Hadrian's Wall in the far north of England. Like many combat veterans before me, I learned that I could not simply outpace the echoes of conflict. They linger, refusing to be left behind.

I embarked on my journey to the Wall in June of 2010, only a few weeks after returning from Afghanistan's Pech River Valley, a place as rugged and unyielding as the memories I brought home. I've never stopped hiking since. Each journey feels like a continuation of something I don't fully understand, a personal odyssey with no end in sight. Perhaps this is the path all who have known war must tread—a road we will walk for the rest of our lives.

There is a truth I've unearthed along this endless trail. It's not about finding the person we were before. That person is gone, irretrievable. Instead, it's about uncovering who we are now—after the fire, after the ashes. The search for this new self continues. For me. For you. For all of us who carry the weight of combat in the recesses of our hearts.

And so, I walk. And I hope.

WE ALL LEAVE WAR, BUT WAR NEVER LEAVES US

Soldiers never forget their time at war. They will always hear the horrifying whizzes and whip-snaps of incoming bullets that split the air around their ears. Their ears continue to ring long after they leave the battlefield. Our time at war is both the worst and, paradoxically, the best time in our lives. The raw, unbearable reality of war strips us bare and leaves us longing for the simplicity within its chaos. Sometimes, we wish we could return to feeling the sharp edge of life and death once more. But we dare not express this aloud. Those who have experienced combat rarely share these thoughts with their loved ones or civilian friends. We hide it, burying the truth deep, fearing their judgment, fearing they will see us as irreparably broken.

Warriors are trained to endure the mortal tests of combat, to face the enemy, and to wield death with precision. We spend months and years mastering our mortal profession of arms. Yet, amidst this preparation, we rarely hear about one of war's greatest dangers—how addictive war can become. No one tells us that the battles we fight will echo long after we return home. The thirst for war, and its lingering hangover, is eternal.

After a year or so spent in the fires of combat, we prepare to leave our battlefields. During this period, we begin our relief in place and hand over our battlefields to new units and soldiers. We pack up our armor, weapons, and gear, preparing for the long journey back to a world that feels foreign—this land of peace. We seal our tools of war inside duffle bags and Connex boxes, leaving behind our combat outposts (COPs) and forward operating bases (FOBs). We then travel to huge bases and scorching airfields, boarding massive planes that promise to carry us away from the domain of death. As the ramp of our C-130s or C-17s closes, we bid our war zones a final farewell.

Yet, the horrid stains and stench of war are difficult to scrub off. The scent of battle, the grit of sand, the indelible shadows of our actions—they never fade. War embeds itself deep within us, carving its mark into our souls. It whispers dark truths, the ones we uncovered in the crucible of combat—truths that compel us to wrestle with a painful paradox: as horrific as war is, we have never felt more alive than when its chaos consumed us.

Some warriors drift like lost ships in the ocean of peacetime. They at times roll aimlessly upon the waves and tides of life. And so, we carry on, like specters of war wherever we go. We move like the shadows on the forest floor. Some men who have survived their encounter with Death are trapped. We are stuck in that place between life and death.

The hunting of man impacts every warrior differently. For some of us, the weight of battle is an anchor we can never shake off. Many of us find that our experiences in combat become haunting specters, inescapable and relentless. We may learn to wear business suits and build lives with families in suburban neighborhoods. We reside far removed from the deafening roars of machine guns and the heart-stopping blasts of artillery fire. Yet, no matter how far we run, the war within us persists, lurking like a haunting shadow that refuses to be cast away.

War's shadow changes form throughout a veteran's life. Now, as I write, well over a decade has passed since I last heard Death's whisper in my ear. Yet the heartbeat of combat still resonates in my head. Various rituals help numb the memories. Exercise, hiking, and writing help, but I remain stained even now. We all carry the scent of war with us. There is hope, and we must try to go on.

This is one story of that effort.

CHAPTER TWO

COMING AND GOING: 1998 – JUNE 2010

I had fantasized about going to war for as long as I could remember. Since watching the original *Spartacus* and *Aliens* films as a child in the 1980s, my imagination had been captivated by battle — the courage, the chaos, the glory. Throughout my youth, I sought to replicate the violence of war through contact sports such as wrestling, karate, and football, testing my limits and embracing the grit. These early tastes of physicality and controlled aggression, I would later realize, were invaluable in preparing me for my ultimate calling: serving in the US Armed Forces.

I've worn a uniform since I was 18 years old. In 1998, I enlisted in the United States Marine Corps and served on active duty for four years, until 2002. My years as a Marine took me to far-off places like Okinawa, Japan, Thailand, and San Diego, where I worked as a topographic intelligence analyst and taught martial arts as a Marine Corps Martial Arts instructor trainer. However, despite my preparation, dedication, and longing for the crucible of combat, my time in the Corps passed without me ever setting foot on a battlefield.

In July 2002, after completing my enlistment, I returned home to Michigan with a renewed purpose: to become a high school social studies teacher. I enrolled at Eastern Michigan University, but something was missing. The allure of military life remained too strong to resist, so I joined the Marine Reserves. When the U.S. invaded Iraq in February 2003, I thought my chance had finally come. I was mobilized, eager to fulfill the dream I had carried since childhood. Yet, once again, war eluded me. Instead of being deployed overseas, I was sent to San Diego and Virginia Beach, far from the frontlines I had envisioned.

In early 2004, I returned to my studies at Eastern Michigan University, thinking I would lead an ordinary life as a teacher in Southeastern Michigan. Although I was a veteran and honored to have served my country in uniform, something was missing from my military journey. I had never had the chance to go to war. For months, I wrestled with this notion and tried to suppress my desire for combat. Yet my hunger for war continued to burn.

This hunger only intensified as 21st-century video platforms, particularly the nascent YouTube.com, brought combat footage from Iraq and Afghanistan directly before my eyes over the Internet. I was engrossed in watching grainy videos of infantry squads trading lead with insurgents in deserts, villages, and mountains. I again wondered what it would be like to go to war. I imagined what it felt like to be in a battle.

This was my generation's war, and it seemed that I'd missed my shot. My enlistment was spent in the gym and experiencing the nightlife of Southern California. Now I was stuck inside college classrooms instead of holding a rifle and fighting my country's enemies. I thought I was missing the only chance of my lifetime to go to war.

I contacted the local Marine Officer Selection Office and started my application process. But an unexpected opportunity emerged. The local Army ROTC detachment offered me a deal I couldn't refuse: they would pay for my master's degree in history. It wasn't the Marines, and it wasn't the immediate path to war I had envisioned, but it was a new doorway—a way to bridge my passion for learning with my unshakable desire to serve in combat.

In 2005, I joined Army ROTC, which allowed me to train as an officer while completing my master's degree. It was a step closer to the front lines. In 2007, I was commissioned as an infantry officer and began my journey toward the maelstrom of combat. After a grueling year of training at Fort Benning, Georgia, I was finally ready to join my unit.

I arrived at Fort Carson, Colorado, in August 2008 and reported to Dagger Company, 2-12 Infantry, 4th Brigade Combat Team, 4th Infantry Division. In May of 2009, we deployed to the Pech River Valley, Kunar Province, Afghanistan. Little did I know that this deployment would define me in ways I could never have imagined.

In Afghanistan, I found exactly what I had been searching for. I had the honor—and the absolute privilege—of leading an infantry platoon in combat. It was a dream, albeit a deadly one, come true. The cacophony of AK-47s, PKMs, RPGs, mortars, rockets, and even IEDs became the brutal symphony of my life. The questions that had plagued me for years—what it was like to face war, to lead men in battle—were answered in blood and fire.

That year was among the best and worst years of my life. It tested me, broke parts of me, and left scars that still linger. Yet, that story belongs to another chapter, already written in my first book, *Siren's Song: The Allure of War.* By June 2010, I was counting down the final days of my deployment, preparing to leave the chaos and return home. However, the truth is that you never completely leave war.

I tried to become normal again. This process takes time, but we all hold on to hope.

This book tells the story of how we, as combat veterans, carry in the deepest corners of our hearts the unending struggle of leaving war behind and trying to reintegrate into the fabric of everyday society.

Each warrior embarks on their journey shaped by the unique battles fought on and off the battlefield. Many of you reading this are likely still navigating that winding path, as I am. Maybe this is a journey we will walk forever, in that space between the echoes of wartime and civilian life.

The simple truth is that the shadows of war never truly leave us. They linger, casting their faint but constant presence over our days and nights. While we can't banish them, we can learn how to live alongside them, to grow old with them, and to somehow find meaning in the life that remains.

This is but one tale among countless others of finding peace in a world that feels alien to the soul of a soldier...

FINAL DAYS AT WAR: FOB BLESSING, PECH RIVER VALLEY, KUNAR PROVINCE, AFGHANISTAN, SPRING 2010

I deployed to Afghanistan in May 2009, serving as a Platoon Leader for an infantry platoon at Combat Outpost Honaker Miracle, just west of the city of Asadabad in the rugged terrain of Kunar Province's Pech River Valley. For eight relentless months, my platoon and I collided with the enemy over one hundred and fifty times—clashes defined by the crack of small arms, the hiss of RPGs, the thunder of PKMs, mortars, and rockets. Each day, we hunted along the single paved road winding through the Pech Valley and ventured into the unforgiving terrain of the Watapor Valley.

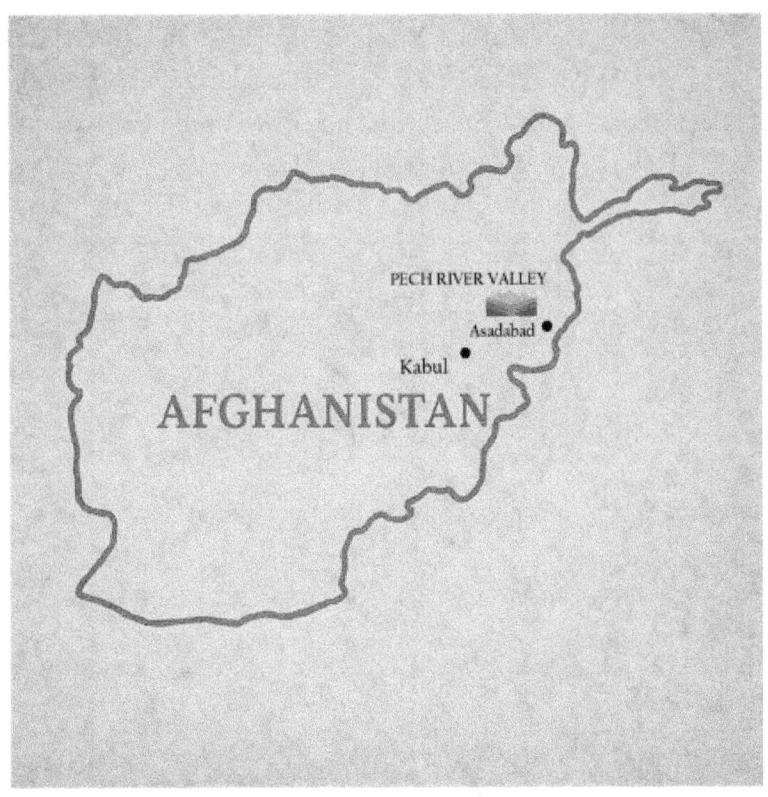

The Pech River Valley, near Asadabad, Afghanistan.

Some days, we endured firefight after firefight, barely catching our breath between battles. I carried the burden of command, leading in combat until the moment came to transition into the role of an intelligence officer.

In late December of 2009, I began the complex process of handing over my responsibilities to a very talented and brave officer, Florent Groberg. Flo, who would later be awarded the Congressional Medal of Honor, took up the reins of leadership alongside my steadfast Platoon Sergeant, Korey Staley. With these two men at the helm, I knew my platoon was in capable hands. Yet, leaving them behind was one of the hardest things I have ever done. I will forever carry the memories of my brothers such as John Wade, Ryan Rojas, Tyrel Richardson, John Moffet, William Stacey, Martin Cortez, Alexander Jones, Allen Mendiola, and Andrew Troxel. These men were my family, my warriors, and leaving them felt like abandoning a part of myself.

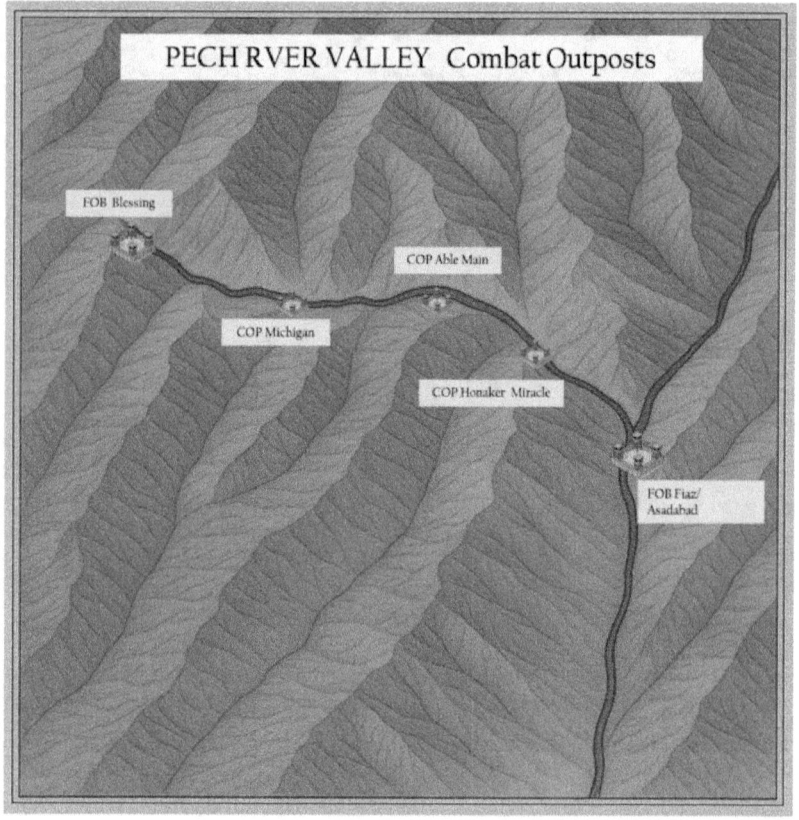

Map of the Pech River Valley

In late January of 2010, I reported to Forward Operating Base (FOB) Blessing, where I spent the last few months of my deployment free of the responsibility of leading an infantry platoon in combat. I worked as an intelligence officer. Or, rather, I *tried* to work as an intelligence officer. The hours of admittedly important work spent reviewing reports, maps, and PowerPoint slides weren't satisfying to me. It was hard for me to give up the near-daily firefights, near misses, and fire missions. I desperately missed the fighting. I missed the rush of battle and the addictive, combat-laced adrenaline pumping through my veins and accelerating my breath. I missed feeling as alive as I had ever been.

I volunteered for patrols whenever possible to dull the ache of separation from that life. Those stolen moments out in the field, rifle in hand, gave me a fleeting taste of the life I had left behind. Even as my final months in Afghanistan ticked by, I knew I was already missing the fight. I was already missing the war.

LAST FIREFIGHT

Sometimes I got exactly what I wanted: just one more firefight. Late in my deployment, I found myself chasing adrenaline-laced combat: the world's most addictive substance. After spending a whole day out on the line, maybe exchanging fire with the enemy, I'd return to the battalion's TOC (Tactical Operations Center) where my shift as an intelligence officer awaited me. Sometimes, I'd walk back into the TOC yelling: "Hell yeah!" still buzzing from the adrenaline of the firefight. I didn't think anything of it at the time. Before starting my shift, I didn't believe there was anything wrong with looking for a gunfight. At the time, it felt like I was going to the gym to get a quick pump in before I went to work.

"WHAT ARE YOU GOING TO DO WHEN YOU GET HOME?"
MAY 2010, PECH RIVER VALLEY

"Salinas! LET'S GO!" Our patrol halted, and I dismounted to the dreadful song of gunfire echoing in the valley. "SNAP! SNAP! WHIZ!. KA-KA-KA-KA! KA-KA-KA!" Sgt. Creed dismounted and began hammering the highlands with his M4. I did the same.

I moved to the corner of the MATV (Mine-Resistant Ambush Protected All-Terrain Vehicle) and, with a click, switched the selector on my rifle from safe to semi. Hearing this click changes the world for a warrior. With this click, I became what I was trained for: Death and the destroyer of men; more than human; imbued with the ability to slaughter my enemies, ripping them apart like a lion shredding hyenas, able to hurl small pieces of lead and rupture internal organs, to rob my enemy of his life and breath. We killed not out of pleasure, but rather just to live. We fought to save not just our own lives, but also those of our brothers and sisters who formed the line of battle.

I was not a platoon leader in this battle, so I did not have to maintain my usual officer reserve of calm, enabling me to talk on the radio, maneuver my troops, or call in an airstrike. Instead, I became a rifleman and warrior of the line. I hammered the highlands.

The enemy appeared in the mountains, not as men, but rather as sparks upon the mountain, visible only as the muzzle flashes of their rifles and machine guns. Their bullets slammed into the road just in front of me. "GO TO HELL YOU BASTARDS!" I screamed at the top of my lungs.

I blasted, repeatedly pulling the trigger, unleashing hot lead. I dropped a magazine and replaced it. After my third magazine, I slowed my rate of fire.

I continued squeezing round, after round, after round. The smoke slowly rose from my barrel until it stung my nostrils. I watched the smoke rise through my ACOG (Advanced Combat Optical Gun-sight-scope). There was something about it — it looked different than it had in the dozens of other engagements I had participated in. Perhaps, that time, rather than the powder propelling my bullets, it was parts of my soul that were burning — the parts I had traded to reap the souls of others.

I left the cover provided by the MATV's armor and slowly walked toward the next vehicle in the convoy. I kept up my fire. Round after round, ejected from my rifle. "Cling, Cling, Cling, Cling, Cling, Cling." The falling brass echoed off the blacktop of the paved road. My ears bled with the ringing of that song of war. For some reason, I didn't feel the fear of battle that I usually did. I didn't even bother taking a knee while switching mags. I stuffed the empty ones into my cargo pocket and continued to hammer the highlands.

I kept moving ... but it all seemed in slow motion. I began seeing bullets strike within ten meters of me. I couldn't move faster... and the impacts came closer and closer.

I felt a force collide with me.

It was not a bullet.

"Sir...! What the hell?!" It was Sgt Johnson. "Are you trying to win a medal? You're a crazy fucker aren't ya?"

Johnson pushed me back into a ditch. "Are you good, sir?" he asked.

"Yeah .... I'm good..." I answered.

"Ok ... sir... we got Apaches inbound. Let's hold here."

"Ok ... I'm good..." I said again, my chest rising.

It was not my time.

We mounted up in our MATVs and headed back for FOB Blessing. We hadn't taken any casualties that day, so we exchanged yells and high-fives. It was a good way to help with the shakes I sometimes got after shedding so much adrenaline. After ten minutes of driving, we returned to the relative safety of our FOB. Unlike the other soldiers of that patrol, who would be going off duty for some rest, I had to return to my actual job as an intelligence officer in the TOC.

Only about 25 minutes after the firefight, I walked into the TOC to begin my watch. I stowed my rifle on the rifle rack by the door, its barrel still warm and carrying the faint smell of gunpowder. My armor came off next, heavy with sweat and dust, leaving me feeling exposed but lighter. I stopped and stared at my workstation — the clean, rolling office chair in front of a sterile monitor. It felt so out of place here, in a world filled with gunfire and grit. The air around me was sharp with the smell of sweat, grease, gunpowder, and tobacco — a cocktail of war that clung to my skin and uniform. I hesitated. My body might have been ready to sit, but my mind wasn't.

"You good, sir?" Sergeant Schlaiss, my sergeant, asked me.

"Yeah… just coming down from the firefight I was just in," I replied, my voice still uneven.

"How'd it go?" he asked, his tone casual but curious.

"It was fucking awesome," I said, a grin breaking through my exhaustion. "I miss being on the line. I really miss getting in firefights all the time," I said.

His expression shifted, his eyebrows raised, a flicker of concern crossing his face. "You really like fighting that much?" he asked quietly.

Schlaiss was good at his job — a sharp, fit, and capable NCO. We had spent hours sparring in the FOB's gym small, worn mats, locking each other in Jiu-Jitsu holds, testing our strength and endurance. But as I looked into his eyes, I saw something I recognized immediately: innocence. Not the innocence of ignorance — he had been in his share of danger here, taken incoming fire, and dodged near misses. But he hadn't crossed that threshold, hadn't stepped into the raw, unforgiving world of life and death that comes with command or giving the order. His brown eyes still held that unscarred humanity.

"What are you going to do when you go home?" he asked, breaking the silence.

"Pssh... I'll be good," I said, brushing it off with a wave of my hand. "I'm good," I said again, the words lingering in the air, sounding less convincing with each repetition.

LEAVING THE VALLEY... MAY 2010

My orders to leave the Pech River Valley arrived in late May of 2010. Even then, I was an avid journaler, spending sunsets after dinner scribbling in my battered Moleskine notebook. I sat on a low wall overlooking what served as our chapel, watching the final rays of the sun drape the valley in dusky rose and deepening blue. The savage beauty of these mountains never failed to take my breath away, even after all the war they had witnessed. This would be my last sunset in the Pech.

By chance, one of my old friends from the Infantry Basic Officer Leader Course, Tim Weed, walked by, leading his patrol from the 101st Airborne. Seeing him brought back memories of those days in the swamps, forests, and mock-urban towns of Fort Benning, Georgia — the long, hot hours spent staying awake during agonizingly dry lectures on operations orders inside Building 4, or what we affectionately called "Building Snore." Those days felt impossibly distant now.

Tim's platoon moved in disciplined silence, a single-file line of shadows, their night-vision goggles mounted on their helmets like unblinking eyes. As they passed, I raised a hand in farewell, wishing them luck, as if that would be enough. They stepped beyond the wire and began their slow ascent into the mountains.

I remained, watching them disappear, swallowed by the terrain, their movements blending with the fading light. They reminded me of waves rolling onto a twilight shore — silent, inevitable, vanishing into the vastness beyond. I decided I would stay until I could no longer see them. This would be my last patrol, even if I wasn't walking it.

The Pech Valley, Kunar Province Afghanistan.

They climbed higher and higher, the last slivers of reddish sunlight draining from the valley as though the land were bleeding out. Then they were gone, disappeared over the ridgeline into the wild, hostile dark of Kunar Province. A warm spring wind curled through my hair, carrying with it the distant scent of Afghan stew, thick with spice and memory.

I allowed my eyes to linger on the valley one last time. The mountains stood as unmoving, silent memorials to our battles, our dead and theirs. I thought back to my first morning at FOB Blessing, nearly a year earlier—when I had woken early, stepped out of our cramped B-Hut, and looked upon these very same mountains for the first time. I remember whispering to myself: *"If this place is as dangerous as it is beautiful, we are in for one hell of a fight."* I had been right.

The sun retreated, surrendering the sky to the deep, black cloak of night that covered the Afghan countryside. The land lay silent, except for the occasional whisper of the wind that gently kissed my face — a soft, fleeting farewell. I never imagined I would feel sadness leaving this place, this valley that had been my crucible. But I did.

Here, in this rugged stretch of the Pech River Valley, I had become a warrior. Here, I had encountered Death, face to face, time and again. And now, as I stood on the brink of departure, I found myself uncertain of the world waiting for me beyond these mountains. Home was no longer the place I had left — it was now a mystery, and I would be a stranger returning to it.

I exhaled, long and slow, as if trying to empty my chest of the weight pressing down on it. Then, I slipped my journal into my pocket and made my way to my hooch for one final, restless night of sleep. I was set to leave FOB Blessing just before dawn.

Sleep eluded me. I tossed and turned, stared at the ceiling, and listened to the sounds of the night — the distant murmurs of soldiers on watch and the occasional rustle of the wind through the concertina wire. Eventually, I surrendered. Dressing slowly, I strapped on my armor, and with a heart caught between relief and hesitation, I began the short walk to the HLZ — the humble helipad that would serve as my gateway back to the world I had left behind.

The moon hung full and heavy in the sky, casting a pale silver glow over this American fortress's steep, winding paths in the mountains. As I reached the makeshift "passenger terminal" — which always reminded me of a barn or an extended car garage — I found a small group of officers and soldiers already gathered. Their energy was unmistakable. The anticipation of leaving war clung to the air like electricity before a storm. We sat together, trading jokes and stories, filling the space with the scent of cigarettes, the sharp bite of energy drinks — Rip-Its, our fuel of choice — and the undercurrent of barely contained excitement.

I still couldn't believe it. After one of the longest yet fastest years of my life, I was leaving. I had survived. There had been so many moments when I thought I wouldn't. Honestly, I believed I would meet my end in Afghanistan—I had even dreamed that I died in a firefight back in 2008. You're not supposed to die in a dream, right?

DREAM WHILE DRIVING FROM FORT BENNING GEORGIA TO FORT CARSON COLORADO – AUGUST 2008

We were on a final assault.

"Move! Move!" I shouted.

We're almost there. We've pushed them back, taken ground inch by inch, and now—just one final surge. The hill will be ours. I know it.

I have to order this push. There's no other way. I have to see it through.

My bayonet is fixed—I'm down to my last magazine. It may come to this. The tool I've trained with endlessly, the cold steel I've driven into dummies on assault courses and sparred with in Marine Corps martial arts. Now, it may finally find its purpose.

I stared at the blade. This primitive weapon—once a spear, a dagger, a sword. The most ancient of human tools. Thrust this into a man, take his life, and live.

Even in the dream, I felt the dread—that primal voice whispering, "Stay down. Stay the hell down."

"The hell with it," I said. "Follow me!"

We rose in our heavy armor, our boots pounding the earth. Fire erupted as my men charged forward, cutting down the last defenders. The hill is so close now.

I sprinted toward a boulder—I knew he was there. Another enemy.

I flicked my selector switch from safe to fire—click.

Nothing. My rifle was jammed.

There was no time to clear it.

He raised his weapon—bang!

I parried his rifle and drove the buttstock of mine into the side of his skull. Crunch. That perfect impact—like blindsiding someone in football, like landing a clean hook in a fight. He crumpled.

And then I stabbed. And stabbed. And stabbed.

He was dead.

But then—I felt it.

The pain.

That bang—his rifle. I wasn't fast enough.

I staggered, praying that it was just the impact against my armored chest plate. I fumbled under my armor, and I felt warm blood oozing from the wound.

The bullet went through.

My legs felt weak, and my body betrayed me. I dropped to my knees, my hands slick with red.

I couldn't stand, and I collapsed onto my back.

"So this is how I go," I thought. I didn't want to go, not yet.

I stared at the sky—a perfect, vivid blue. I remembered this same sky at boot camp, where I stared up while doing flutter kicks and gasped for air. I felt sad knowing that I was going to die.

I woke up from that dream before heading off to war. I had just graduated from infantry school and was driving from Ft. Benning. I found myself in the middle of Kansas, *en route* to Ft. Carson, CO. Drenched in a cold sweat, I stared at the alarm clock; it read 2:12 am. This was particularly alarming because I was scheduled to report to my unit, the 2nd Battalion, 12th Infantry Regiment—2-12 IN. I recalled the play *Julius Caesar* and the omens Caesar's wife, Calpurnia, experienced in her dreams on the eve of his assassination. I was convinced that I would die in the war.

Ever since that dream—long before I heard a bullet crack in combat—I had carried a quiet certainty that I wouldn't make it home. Yet, here I was. As the eastern horizon began to blush with the first pale hints of morning, a distant hum whispered across the valley. It grew louder, deeper, more rhythmic—the unmistakable thudding heartbeat of rotor blades cutting through the air. To a soldier, that sound is music. In battle, it means salvation. The heavy thrum of approaching helicopters always signified something—survival, reinforcements, or rescue.

Apache helicopters came with their talons full of Hellfire missiles and 30mm vengeance. Kiowas with their .50-cal machine guns and rockets. Blackhawks, angels in the sky, carrying the wounded home or delivering the lifeblood of ammunition and water. But today, it was the Chinook helicopters.

Their approach transformed the hum into a deep, pulsing *thud-thud-thud* that resonated in my bones. These were the same type of heavy-lift helicopters that had carried me into this valley in May of 2009. Now, they had come to take me away. Dust and sand filled the air as they descended, coating my skin, filling my nostrils, slipping into my mouth—giving me one final taste of this war-torn land. I stood there inhaling it, letting it settle deep inside me, as if some part of me needed to hold onto it and carry it home.

The prop wash roared against my face, warm and relentless, as the ramp lowered. A crew chief leaned out, signaling. "Okay, LOAD UP!"

This was it.

I took one last look at the valley—the tall mountains, silent witnesses to all we had done—all we had lost and taken. Then I stepped forward into the waiting bird and the unknown.

I have always found the fully helmeted helicopter crew members to be an almost alien presence. Their faces hidden behind visors, their heads encased in heavy helmets, they resembled stormtroopers from Star Wars more than men of flesh and blood. One of them stood at the ramp of the Chinook, his gloved hand motioning for us to come forward, inviting us to step aboard his machine and leave this war behind.

The rotor blades pulsed with an eerie glow, static electricity dancing along their edges like ghostly tendrils in the night. The ramp—a cold sheet of steel manufactured in some distant factory by men who had never seen this land—touched the dirt of the ancient battlefield, a place that had swallowed countless invaders before us. I trudged forward with a rucksack slung over my shoulder and a duffle bag clutched tightly. I hesitated for the briefest of moments at the threshold, pausing before placing my foot on the ramp, ensuring I wouldn't trip and fall backward into the valley and onto the Afghan dirt.

I settled into the Chinook's basic webbed seating. A simple nylon bench stretched across the interior, a stark contrast to the armored war machine that surrounded us. I sat in silence, waiting. I was waiting to be lifted away from this land where I had fought, killed, ordered men to kill, and watched good Soldiers die. I was waiting to be carried back to a world of peace—a world I was no longer sure I belonged to.

More men climbed aboard, each carrying their own ghosts and burdens. A silent prayer slipped through my lips as we sat there, vulnerable inside this metal shell soaring through the sky. *Please, no RPGs. Don't let it end like this.* But we all knew the pilots wouldn't linger. They were as eager as we were to leave.

The Chinook lifted, the earth falling away beneath us. I exhaled, a breath I felt like I had been holding for a year. The world outside shifted to that deep, ethereal shade of blue that every soldier and hunter knows—that pre-dawn hue where the sky is neither night nor day, a liminal moment between worlds. We were leaving, rising above FOB Blessing, heading east. I had only just begun the journey of leaving war.

I watched the mountains through the small window beside me, and I swore they watched me back. These peaks knew things—secrets of war, of men, of life and death. They had absorbed my sweat as I climbed them under the crushing weight of armor. They had echoed with the sharp commands I shouted in battle. They had taken in the thunder of our rifles, the crack of enemy fire, and the screams that followed. They understood my fear, my rage, my exhaustion. They knew all of me.

The Pech River glowed silver in the dawn, winding through the valley like a serpent. I recognized the landmarks instantly—COP Michigan, standing as the gate to the Korengal Valley. There, I had first heard bullets snap past my head and first returned fire, not with my rifle, but by hanging the heavy 120mm mortar rounds alongside my friend Mark Zambarda. Further east, we passed COP Able Main, and I recalled the dozens of firefights fought in its defense. Then came COP Honaker Miracle, my home for most of the deployment. I looked down at the villages, the ridgelines, the narrow draws—battlefields I could name as if they were streets from my childhood.

And then, unexpectedly, a lump formed in my throat. I felt the tears before I could stop them. As we crested the final ridge, leaving the Pech River Valley behind, something inside me cracked. My war was ending. My year of surviving, of fighting, of killing—it was over. Even as the Chinook turned south towards the Kunar River Valley, I knew that, even as I was leaving the valley, the valley would never leave me.

The Chinooks flew us to Jalalabad Airfield—JAF. There, the war began to fade. It was quiet. Almost too quiet. No incoming mortars. No gunfire. No frantic radio calls for air support. Just the hum of transport planes and the steady rhythm of troops cycling in and out. No artillery. No bombs. Just the dull roar of engines from C-130s and helicopters.

I tried to stay busy helping with paperwork, lifting weights, and counting the hours. But time moved strangely here, stretched thin like a ghost of the war I had just left. I was no longer in battle but wasn't home yet either.

Then, one night my name was called. It was time to go. We walked onto the tarmac beneath a sky so clear and bright it seemed painted. The full moon bathed everything in silver, stretching our shadows long and dark against the pavement. I looked around at my fellow soldiers, their helmets and rifles silhouetted in the night. We glided like ghosts across the airfield toward our waiting C-130. We strapped in and I settled into my seat for my final flight in Afghanistan.

Bagram was next—the gateway out of this war. The vast base, teeming with troops and firepower, lay beneath the towering peaks of the Hindu Kush rising in the distance. Those mountains had been here long before us, and they would remain long after we were gone. They had seen invaders come and go and had witnessed ambitions rise and fall. And as I sat there, watching them fade into the distance, I wondered if we had ever truly understood their silent warning:

Some places cannot be conquered. Some places should be left alone.

The time had finally arrived to pass back through this gateway and head back toward the land of the living. I took in my final Afghan sunset near the Bagram airfield and watched as the sun bathed the mountains in a glow of orange light. Darkness soon covered the land, and slowly, the full moon illuminated the land with a silent glow. I stood there with a few other officers I served with, such as Greg Sullivan. I stared at our shadows on Afghan soil for the last time. I walked aboard the C-17 Globemaster ramp, set my gear down, and took my seat. As the ramp to the mighty bird closed, I gazed at the dark mountains illuminated by the moonlight one last time.

In Afghanistan, I finally found what I had searched for my entire life: war. I had escaped the war with my life. My physical cost was not too dear, besides some minor wounds in an IED blast. But the cost to my mind and soul was more treacherous and would take much longer to heal.

I sighed as the ramp slowly closed, and I bid farewell to many things in those last few moments. The Hindu Kush Mountain range bid itself goodbye to me. I said farewell to those mountains and the terrors in her shadows and valleys. I said farewell to my war. However, years later, I realized that I was saying goodbye to much more.

A part of me was left there and never returned. I bid it farewell.

We became airborne and made the short flight from Afghanistan to Manas, Kyrgyzstan. Here, we spent a few more days on that strange odyssey toward home. This place reminded me of stopping in Kuwait during my mid-tour leave. It felt like one of those passages between the dead and the living. This airbase was undeniably one of the crossroads of existence. There were soldiers from several NATO countries. I walked past the UK forces with a hint of jealousy, as they were permitted to drink.

Amongst the many faces and units, there were also incoming troops from the 101st Airborne Division, the unit scheduled to replace us. I looked at their faces and noticed a difference between the seasoned veterans who understood what awaited them in combat and the fresh faces of the young officers and soldiers who had yet to experience the horrors of war. They were moving into the storm we had just left. Clearly, the war was no longer ours—it was theirs.

We continued packing for the voyage home. It became increasingly clear that we were headed for peace, as we were told to stow away our armor in our bags, no longer needing it. I gazed down at my dirty IOTV (Improved Outer Tactical Vest) and carefully opened it to remove the heavy Kevlar plates. These pieces of armor were made to protect our torsos from shrapnel or bullets, particularly our vital organs. It felt strange as I kept taking them out of my sweat-and-mud-stained armor. We sealed them up, storing them in our bags. I packed up more and more of my warrior self until it would be needed again—if it ever would.

Beyond our armor, we also packed away our brown Multi-Cam camouflage uniforms, as we had been issued brand-new uniforms for the flight home. I unpacked my new uniform, the Army Combat

Uniform (ACU), and stared at fabric that had never seen action. It smelled clean and pure, and even felt soft and innocent. I easily repacked my old combat uniforms. Shedding the weight of war would not be nearly as easy.

The day finally came to head back to the USA. We made one more stop in Frankfurt, Germany. We were kept separate from the rest of the airport in our terminal while waiting for our final flight to the United States. We were permitted to go outside to a small open-air break area, and we were greeted by green grass and dandelions under a pure blue sky. I calmly chatted with some soldiers as we prepared to return to the land of peace. We ate the "German" sausages they had for sale in our contracted part of the airport and waited. There were no sounds of gunfire or explosions, as we were on free NATO-held ground. Soon, we boarded a commercial flight back home to the United States.

US AIRSPACE

I'm coming home.

While I felt physically comfortable, I felt out of place on a civilian plane. The air conditioning, bags of peanuts, and cans of Coke served by young women felt strange and foreign. I tried to relax as I adjusted to an environment free of fear, dust, and heat. I wore no Kevlar armor, just my gray digital camouflage ACUs. We even had to stop wearing our rugged civilian hiking boots and switch to the standard Army tan boots. I was so used to managing multiple tasks from the commander's chair in my truck or the operational center that it was hard to sit comfortably in my chair and do nothing.

There was no radio chatter, machine gun fire, or distant booms. Just the constant hum of the airplane's engine. *No longer* did I make sure that I had fully loaded magazines in my kit, with my NODs, batteries, and water. *No longer* did I have to scan the highlands for enemy movement. *No longer* did I have to check to see if my anti-IED device was constantly on. *No longer,* did I have to look at my blue force tracker (command) screen and take note of my grid. *No longer* did I have to look outside my window and see which artillery target was closest to call in if we got hit. *No longer* did I have to check in with the nearest combat outpost and tell them to dial in their mortars to support us just in case. *No longer* did I have to monitor my vehicles to ensure our spacing was too close or make sure the men did not bunch up on patrol while on foot.

There were many things that I no longer had to do. But what was I supposed to do now? Who was I supposed to be in this world of peace? It was almost time to return to the world where we no longer existed to serve in battle, where we no longer hunted man. Perhaps that world would be the same. But I was not.

Sleep eventually found me. The offer of food occasionally roused me, and I finally stirred while flying over the Western United States, near Colorado. It was early June, and as I looked out the window, the deep blue of the Colorado sky welcomed me. In the distance, the Rockies rose majestically like titanic guardians of the American frontier. I shared a few last jokes with my fellow soldiers sitting nearby.

We began our descent, and everyone turned quiet.

We were in the final moments of our war.

I prepared to enter my peace.

My elation at surviving my first combat tour began to deflate as the houses outside became visible from my window seat. A mixture of fear, discomfort, and stress started to churn in my stomach. Divorces had become all too common in the deployment environment during the height of the Global War on Terror. Many marriages had dissolved or were in the process of dissolving during that deployment, and my marriage was no exception. There would be no house with a "Welcome Home" sign waiting for me. I would return to no one.

We landed at Colorado Springs Airport, where the 4th Infantry Division band welcomed us by playing the Rocky theme as we deplaned, which made me smile. Various high-ranking colonels and generals greeted us on the tarmac, smiling, thanking us, and shaking our hands. I locked eyes with one of these seasoned officers, whose rank and name escape me. While he smiled, I recognized in his eyes the heavy burden of grief that accompanies all command.

After deplaning, we took a short walk to a nearby hangar. Immediately, we formed lines to turn in our rifles. I used those final moments to bid farewell to my rifle. I held it tightly as I waited. This weapon had been with me for the last 13 months at war, and it felt like I was saying goodbye to a loved one.

In battle, I kept her close when the unforgiving, blood-curdling snaps of incoming bullets pierced the air around me. I had her and only her when the all-consuming black smoke and fire of an IED blast enveloped my vehicle. I held her while I screamed in return fire during firefights. As my turn in line approached, I sadly looked down at her, as the scent of CLP and the bittersweet taste of gunpowder flooded back to me. My fingers traced her pistol grip gently. Even now, I can recall every curve and every grain of sand and dust that lingered, remnants of the savage lands we had just left.

My turn finally came to the front of the line.

I signed her in and bid my farewell.

Goodbye, my love. It is you who knows me all too well. Just you...

* * *

We waited for transportation back to Fort Carson. We all sat down, casually bullshitting like all soldiers do to pass the time. Soon, we were ushered into a large bus that reminded me of the ones used for tours of the Gettysburg battlefield. I sat down, looked to the west, and gazed at the rugged beauty of the Rocky Mountains. While breath-takingly beautiful, the mountains seemed to evoke an emotion I was not prepared for: fear.

For the past year, I knew the enemy owned the mountains, and so did they. We could only rent out space in the highlands, paid for with blood and thousands of dollars of bullets, missiles, and bombs. Every god-forsaken patrol into those peaks seeped our strength. You have not been winded until you have engaged in combat patrols walking up mountains. Every step taxed my breath. Every step told me that I was in an enemy country. Every patrol was a journey into the darkness.

The mountains were always there in Kunar Province, Afghanistan. They watched us, serving as the enemy's watchtowers. From their mountain strongholds, they rained fire upon us with their rifles, their machine guns, and their RPGs, with their mortars, and with their rockets. These mountains were the silent dragons that breathed fire upon us and strove to destroy our bodies.

I used to watch these mountains, even during rest and refit days. I could stare at our battlefields on these unnamed peaks from my combat outpost. I stared at where I could have died. I stared at the places where a part of me *had* died.

But the part that survived was shaped there. There is no forge like that of war.

I reminded myself that no fire would breathe down from these highlands in Colorado. The orange, Mars-like rock at the summit of Pikes Peak welcomed me back. I tried to set my mind at ease. Even today, I love hiking mountains, but at first glance, they dig up those hidden primordial memories of life and death.

The short drive down I-25 was a surreal experience, yet another introduction to this familiar, yet oddly foreign, world. This beautifully maintained and paved expressway was a far cry from the poor excuses for roads that terrifyingly hugged treacherous ravines in Afghanistan.

I tried to remind myself that here, on this road, I could relax. There would be no ambush, IED, or RPGs. I smiled as I stared outside at the flashing siren lights from the police vehicles escorting us, clearing the way ahead and giving us a "hero's" salute.

Soon, we passed through the front gate of Fort Carson, and a collage of peacetime memories spread across my face. I recall my first day as a platoon leader in the fall of 2008 when I was pulled over at 5:00 am for speeding on I-25. I wasn't speeding because I was running late but rather because I was excited to start my first day in that role. The police officer simply smiled when I shared my excitement, and he let me off with a warning.

As our buses lined up at the gate, I remembered the countless times I stood in that line in my truck, hoping I wouldn't be late for the pre-PT company leader meeting at 0600. Our convoy of buses continued and entered Fort Carson. Before long, we approached the fort's Special Events Center, which was essentially a large gymnasium that reconnected families after deployment.

We quickly got off the bus and formed a parade-like marching formation. After a year of patrolling, being in such a formation, and standing so close to each other without at least five meters between me and the guy next to me felt strange and vulnerable. My old company's First Sergeant stood at the front of the formation, reminding us to put on a good show as we marched in. I locked eyes with him and saw that his smile tried to mask his grief and worry for us. He did his job well.

I gathered with the soldiers around me, and we marched in. The gymnasium doors opened, and we were enveloped by a cloud emitted from a smoke machine. The thick synthetic smoke transported me thousands of miles east. It reminded me of the smoke that choked me as I tossed a smoke grenade, marking my first hot HLZ (Helicopter Landing Zone) in the Watapor Valley in July 2009. I realized then that leaving war would not be as simple as taking a flight out of a combat zone. While my body was home, it would take years for my mind and soul to catch up.

HOT HLZ WATAPOR VALLEY, JULY 2009

"CRACK CRACK CRACK… whizz…whiz…." We were pinned down in an irrigation ditch. My entire company was in a fierce battle with the enemy, who poured relentless fire upon us from the high ground.

Me in our platoon's defensive position in the Watapor Valley.
Photo courtesy of Eros Hoagland.

We had several wounded now. The Taliban had his shit together. "FUCK!" My commander shouted and threw the microphone to his radio down. The enemy closed to within 300 meters of our position.

My radioman, SPC Jones, carried the radio. SPC Jones looked at me with blue eyes, which I always thought looked too innocent for an infantryman, and asked, "Sir, are we going to be okay?"

"Yes," I lied.

The truth was I thought I was going to die with that boy in a muddy ditch. Fear resonated from the whizzes of bullets overhead and the horrific screams of our wounded echoing on the radio. You don't know fear until you think you are within inches of death. You think you will be brave, but the self-preservation instinct screams to stay down.

It was only a few heartbeats, but I was a coward for an instant. I wanted nothing more than to hide from the firestorm of bullets and RPGs whose cacophonic orchestra sang around my ears. In training, I was always brave, and the words "follow me" always came easily to my lips. However, once you feel the icy grips of death upon your neck, bravery can sometimes be hard to find. I stared at the muddy water that covered me up to my shins and somehow failed to cool me in the 110-degree heat.

Fear is a terrible thing, but while it is impossible to expunge it, you can learn to deal with it. I licked the salt off my lips and reminded myself that I still had life and a job to do. We had men who desperately needed medical attention, and the entire world appeared to be on fire.

The quick reaction force, led by my battalion commander, LTC Brian Pearl, arrived on the scene and provided much-needed extra firepower. He came to us and his voice helped to end my momentary trance.

"Salinas! Do you have a smoke grenade on you?" he asked me.

"Roger sir!" I answered.

"Good. Now go make an HLZ and get these wounded out of here!"

"Roger that, Sir!"

"Damn! This is not going to be easy." I thought to myself.

The valley was filled with smoke and the terrible sounds of gunfire, mortars, and explosions. I had lost count of how many bullet impacts had landed within a few feet of me, continuing to shatter the sound barrier in the air just above my ears. My body ached. We had been on this mission since sunrise, and it was now mid-afternoon. I was thirsty, but everyone was out of water. I glanced down at my boots to ensure they were tied. I double-knotted them to be safe.

I slapped a fresh magazine into my rifle and looked at Jones and a few other men from the Headquarters section.

"LISTEN UP!" I shouted. "Guys, we are going to be okay!"

I could tell the men were not focused on me, as they eagerly looked back and forth at the surrounding highlands, distracted by the sounds of battle.

"Get your eyes on me!" I screamed. "Do you see that building over there? We are going to haul ass over there and set up an HLZ! We need to be fast. Everyone understand?"

The men all nodded.

"Ok. Follow me!"

The small fire team of men ran under an umbrella of friendly and enemy machine gun fire. I fired a few rounds into the highlands, and my ears rang.

EEEEEEEEEEEEEEEEEEE!!!!!!!!!!!!

My ears caught on fire from the shots of my own rifle. This ringing is the truth of combat. Anyone who has tasted modern battle knows this noise. The ringing drowns the world away from you. You are deaf, yet you are not deaf. You can hear breathing, yells, and more shooting. You are alone in this ocean of chaos. This is Death's song. She talks to us when only we can hear her.

"Fuck!" I screamed as a few bullet impacts barely missed me as I ran. We took cover briefly under a stone wall. The sprint winded each man, and I saw their chests rise and fall under their body armor.

"Bound by buddy teams! We'll cover you! Ready? Go!" We covered them with a wall of lead from our rifles. They made it. I ran and joined them.

We reached a small farm building and took cover behind its ancient mud walls. Behind it was a large terrace with a field to serve as my HLZ. I was happy to have some cover for the wounded as they awaited the helicopter.

Another fire team of troops arrived from the Quick Reaction Force, carrying the wounded men on litters. SSgt Josh Campbell was amongst the men, and I was happy to have his assistance moving the wounded.

I called for the MEDEVAC.

"Dustoff 44, Dagger 46… I have a MEDEVAC request. Line 1…"

I called out the rest of the 9-line MEDEVAC request.

Soon, I heard the Blackhawk's rotor humming closer in the valley. I stared south and saw them, dark angels flying through the heat. I grabbed the smoke grenade from my pouch, popped it, and threw the smoke into the center of the HLZ. I inhaled some of that red smoke, and it choked me. I was so thirsty but still was out of water. We waited. As the Blackhawk descended lower and lower, it emitted a small artificial sandstorm.

"Let's give the bird some cover! Suppressive fire!" I screamed.

The MEDEVAC chopper's rotor wash only ignited the hot Afghan summer air further, sending waves of dirt and sand into my nostrils and mouth.

With its technological might, this beast almost resembled an alien spaceship in this godforsaken valley.

The Blackhawk landed, and the rotor wash felt like a dragon was breathing down upon us. The wounded and we were peppered with dirt and sand. The flight medic ran toward me in the storm of flying debris, like an angel emerging from a storm.

"Are these all the wounded?" he asked.

"Roger.…"

I met the eyes of the wounded men. A few of them lay there with their limbs mangled. One's man's limb hung to his body by only a few tendons. I could see his skeleton under the bandages. The sight was horrifying.

The wounded warrior looked at me.

He raised his hand weakly, and I held it. He squeezed me tightly, as to show me and God that he was not ready to die. His breathing was slow, and the medic who worked on him stared at the ground. These were bad signs.

"Am I going to die?" he asked.

His skin was pale, and his eyes were already dull.

"No. You're going to be fine," I lied in as comforting a voice as I could muster.

Shouting orders on my first HLZ. Photo Courtesy of Eros Hoagland.

* * *

Back at Fort Carson, we marched through the swirling smoke into the vast gymnasium, its towering fluorescent lights casting a sterile glow. The brightness reminded me of my old high school gym—the same kind of lights that illuminated countless wrestling matches from my childhood. The air burst with applause and cheers from the crowd of wives, children, and family members. The noise should have been comforting, but instead, it made me uneasy. I felt grateful to have survived, but I already sensed that my homecoming would be more complicated than I'd imagined.

"Battalion! Halt!"

"Left face!"

"Parade rest!"

We froze in formation as our commander took the stage to give a brief speech. I can't recall a word he said. Instead, I scanned the crowd, searching desperately for a familiar face... but there was none. I already felt adrift in this strange, chaotic peace.

Then—"Dismissed!"

The cheers grew louder, but I barely heard them. It was my turn to face a different kind of battle—finding peace. Around me, the soldiers broke formation and embraced their wives and children. My heart sank as I realized that my marriage was over. We had married just before my deployment, and it hadn't survived the test of combat.

My new apartment wouldn't be ready for a few more days, so I made my way to a hotel. I walked into my room and collapsed onto a bed. I began to cry.

I wept—alone—for the first time since I was a child. The weight of war and divorce, of love lost and innocence shattered, pressed down on me all at once. Memories flooded back, cascading down my face in a waterfall of tears.

I cried myself to sleep.

CHAPTER THREE

AN ATTEMPT AT REINTEGRATION

I awoke alone the next morning for the first time in over a year. I rolled over, disoriented, a stranger in my own body. It took me a few moments to remember where I was. Sitting up, I gazed at the solitary piece of modern art on the hotel room wall. The vibrant colors seemed almost alien to me. Then it struck me — this was the first piece of art I had come across in more than a year.

I sighed. My thoughts drifted to Martin Sheen's character in *Apocalypse Now*, sitting alone in a Saigon hotel room, trapped in a world that no longer made sense. The first time I watched that film, I struggled to understand his restlessness and disconnection. But now I knew. After experiencing combat, some of us never forget how it feels. Some of us remain haunted by it, unable to shake off the burden of war. My own war hangover was just beginning.

A month earlier, while I was still in Afghanistan, I had planned a trip to hike Hadrian's Wall. I had been looking forward to it for weeks, counting down the days until I could escape into that ancient landscape. But that trip was still a few weeks away. I needed something to occupy myself in the meantime — anything.

I drew back the hotel curtains and gazed westward. The Rocky Mountains rose before me, illuminated by the morning sun, their peaks glowing with light. They were nothing like the jagged, treeless ranges of Kunar Province in Afghanistan. These mountains felt distinct. They beckoned to me.

"I need to be back in the mountains," I said aloud.

After a quick breakfast, I headed to REI to prepare for my next adventure. I picked up a good rucksack, a hiking stick, a sleeping mat and bag, a tent, and some clothes. I was ready. One of the best aspects of living in Colorado—particularly Colorado Springs—is the easy access to the Rockies. One of my favorite trails in the world is the renowned Barr Trail, a thirteen-mile ascent to the summit of Pikes Peak, one of Colorado's iconic 14,000-foot mountains. I spent the rest of that day relaxing and preparing for my hike on the following morning.

I set out before dawn for the short drive to Manitou Springs. The sky remained gray as the sun spread its rose-red kisses across the horizon. The cool, clean mountain air brushed against my skin like an old friend. Manitou Springs looked just as I remembered—a town nestled into the Rocky Mountains, deep within their embrace. I parked my truck, slung my pack over my shoulders, and headed toward the trailhead.

As I started my hike, I noticed storm clouds gathering in the east. I probably should have turned back, but I knew I wouldn't—I couldn't. The sound of rocks crunching beneath my feet was familiar and comforting. I pressed onward, climbing the first switchbacks of Barr Trail. Soon, I was sweating, my breath becoming heavier in the thinning air. But my body remembered how to do this. The animal breathing returned to my lungs.

I couldn't help but see the trail before me through an infantryman's eyes. I tried to shut my hunter's mind off, to silence the warrior inside me, to strip the war from my soul for just a few minutes. But I came to realize, after all these years, was a simple, undeniable truth: There is no turning it off. I looked at the ground ahead, but not as I once had. Not as a hiker. Not as a man simply moving through the wilderness. I viewed the earth through a new perspective—predator's eyes.

A civilian's wide, round gaze had vanished, replaced by something sharper, forged in the crucible of war. It felt as if my pupils had narrowed into a tiger's vertical slits, absorbing light, not for beauty or peace, but for survival. Each ray of daylight that struck my retina no longer transmitted images to my brain for mere appreciation. No — these signals were entirely different. They weren't for wonder. They weren't for joy. They were for combat.

I scanned the terrain before me, not for its natural splendor, but for advantage. Each stream bed, outcrop of boulders, shadowed group of trees — these were not just features of the land. They were cover and concealment. They were fighting positions.

I was no longer hiking. I was on patrol again.

A massive BOOM shattered my martial meditation — a thunderclap that echoed through the mountains. For a brief moment, my soul was yanked back into the clutches of death. My heart raced into overdrive, adrenaline flooding my veins in a familiar, electric surge. Instinct took over. I dove for cover behind a boulder just off the trail.

The sound had transported me — thousands of miles away, back to my bloody battlefields. Because that sound meant something else to me now.

BOOM! JULY 2009

The Earth shuddered. My MRAP rocked violently. The very air convulsed with the explosion of an RPG. Then came a sound I will never forget. It wasn't just a boom. It was the epitome of fear and dread — a sound, unlike fireworks, unlike simulated munitions. It carried an unmistakable truth.

Death.

Before that moment, I had wondered what a demon or a dragon from ancient lore might sound like. Now, I knew. The gates of hell had opened in Kunar, and her demons, armed with Soviet-style weaponry, charged toward us—hungry for our souls and eager for our bodies.

Time warped. It stretched, folded, and froze. The sun-drenched valley blurred in a haze, swallowing everything—trees, rocks, and even time itself. I stared at the highlands, struggling to breathe, scanning the jagged terrain. They pulsed with heat and hatred, concealing Death itself. I couldn't move.

My mind raced through a thousand thoughts at once. I had read about moments like this. But feeling it—this was different. Warriors freeze in these moments because their souls touch the shores of the River Styx—one foot still in life, the other already stepping into the unknown.

My feet were wet.

I had touched the other side.

KA-KA-KA-KA! KA-KA-KA-KA! POP-POP-POP!

Gunfire.

The roar of my platoon's weapons ripped me back into my body. Our trucks answered with their own song—.50 cals, 240B machine guns, and the thunderous bark of my MRAP's MK19 (automatic grenade launcher).

"Honaker Miracle, this is Dagger 46. We are receiving RPG and small arms fire at Grid…" My voice came out clear, steady—somehow. I had no idea how I managed to choke the contact report out.

The ground around our vehicles exploded with impacts. Dust kicked up, whispering impending doom. But we hunted back. We answered with a menu of munitions hurled toward the enemy. Our trucks rocked back and forth, keeping them from getting a clean sight picture.

We danced together, the enemy and us. It was a forbidden dance with Death.

I heard deep breathing over the headset's intercom somewhere in the chaos. PFC Cortez—my gunner. He was still firing his MK19, but his breath was ragged and shallow. I heard each breath emitting both fear and survival.

I called him over my headset.

"Cortez! Control your rate of fire. You're going to be fine."

"Roger that, Sir!"

I kept my words short and direct. But I wanted to tell him so much more: "Cortez... I can't promise you that you'll leave this alive. None of us might. Death is here, watching us. But you know what? Fuck it. We'll make them fall instead of us. And if we have to go, let's take a few of them with us."

"Dagger 46, this is COP Honaker Miracle." My outpost called to me.

"Be advised, the Scout Weapons Team (helicopters) is inbound to your location at this time."

* * *

As I lay there in the Colorado dirt, time stretched and distorted, just as it always did in battle. For a moment, all sound faded except for the pounding of my own heart, each beat pumping adrenaline-filled blood. My stomach tightened, and my hands were slick with sweat. I scanned the highlands, searching for an enemy who wasn't there.

I then realized it was only thunder. The world slowly came back into focus. I exhaled and stared at the ground, feeling a strange mix of relief and embarrassment. I had heard stories about guys with PTSD—or whatever they called it—but I never thought it would happen to *me*. That was something that happened to *other* men.

And yet, there I was—lying in the dirt, taking cover on a mountain trail because of thunder.

"Are you okay?" a woman's voice called out.

The voice startled me. I turned to see a woman standing a few meters behind me, looking at me with concern.

"Yeah... the thunder just startled me," I said, pushing myself up and dusting off my pants.

"It was very loud," she offered.

"Yeah... it was."

Her name was Victoria, and I told her I had just returned from war. She smiled—not out of pity, but with an expression that told me she understood. We walked together for about an hour, talking as the mountain air wrapped around us. At one point, we took shelter beneath an overhang as a gentle rain fell. There was something peaceful about the moment. As luck would have it, she was a therapist and very easy to talk to.

For the first time in ages, I felt at ease. When the rain stopped, we went our separate ways. I watched her fade down the trail, then shifted my gaze back up the mountain.

I kept climbing. I hiked on, but my mind stayed on that moment—the way the thunderclap had jolted something deep inside me. I had heard thunder a thousand times before, so why did I jump this time?

I thought about the primordial weight of thunder and lightning, the way they have haunted mankind since the beginning. How, for thousands of years, these sounds had sent people running for shelter. Maybe I wasn't all that different from them—ancient hunter-gatherers, barely noticing the shifting sky as the soft gray overcast darkened, deepened, and swallowed the sun.

The wind stills. The birds go quiet. Then—BOOM. The first crack of warning. No hesitation, no delay. The men move to shelter—not in panic, but with the kind of certainty only survival can teach. They don't need to see what lightning can do. They know to fear it.

That ancient instinct had never left me. The storm threatened, but I wasn't turning back. Nothing awaited me back down the mountain—no one was in the hotel room, and there was no warm homecoming.

So, I continued my ascent. The mountain provided me solitude and peace. The cool shadows of the woodlands stretched across the trail, enveloping me gently as I advanced. My pack was light — only twenty pounds, a fraction of the sixty I had carried during the war. No armor, no ammo, no weapons — just me and the mountain.

By midday, I arrived at the tranquil clearing of Barr Camp, where I sat down to rest and enjoy my lunch. The storm threatened in the distance, but for now, the mountain still welcomed me.

After lunch, I pressed on, determined to reach the summit of Pikes Peak. The combination of thinning air and my jet lag made every step more challenging. As the altitude increased, my struggle intensified. My breath became heavier, and my legs felt weaker. Once I crossed the tree line, fatigue settled into my bones, nearly bringing my pace to a crawl.

I glanced at my watch and realized I was about to miss the last train down from the summit. I probably should have turned back, camped at Barr Camp, and called it a day. But something wouldn't let me. Maybe it was stubbornness, or perhaps something deeper — an instinct to push through, to discover my limits and exceed them. So, I kept climbing.

Despite my exhaustion and minor hypoxia, I enjoyed the peace in the barren, rocky landscape stretching toward the summit. Above the tree line, the world was stripped down to its rawest form — just stone, sky, and silence. Near the top, I passed the memorial plaque for G. Inestine B. Roberts, who died on her 14th ascent of the mountain in 1957 at the age of 88. *"What a way to go,"* I muttered, smiling.

At last, I reached the summit — about thirty minutes too late. The last train had already left, disappearing down the tracks without me. It was a little past 4:00 p.m. I stepped into the café at the summit, grabbed a small meal, and considered my options. The sky was clear, and I had enough supplies to camp if necessary. My original plan had been to make it back to Barr Camp before nightfall and sleep at around 10,000 feet, where the air would be easier to breathe. But then one of the café staff suggested something else.

"You could hike down along the railroad tracks," they said. "It's only about six miles down that way, instead of the full thirteen miles on the Barr Trail."

I was already exhausted, so it seemed like the best option. I took their advice and started my descent along the now-silent tracks. Although the distance was shorter, the steep incline of the railroad was brutal on my already fatigued legs. Each step sent new waves of exhaustion through my body. Then, as the sun dipped below the horizon, the world was suddenly bathed in silver. The full moon rose, illuminating the mountainside with a brilliance that felt almost surreal. I hardly needed my headlamp—the moonlight was sufficient, casting sharp shadows across the terrain. The night was warm, the air still, and I felt truly content for the first time in a long while.

Hiking made me feel like myself again. The fatigue was still present, pressing against me, but it was the kind that calmed my mind instead of setting it racing. There was no worry, past, or future—just the rhythm of my steps and the steady bounce of my shadow on the Earth. And for that moment, that was enough.

But I had let my guard down, eyes on the ground, lost in thought. And, suddenly, a flurry of movement and sound shattered the silence.

A large animal bounded across the railroad tracks just ten yards ahead of me. Instinct took over. My body reacted before my mind caught up—I lunged for cover, sprinting to a nearby powerline post just off the tracks. I turned on my headlamp. The beam sliced through the darkness, landing on an animal, a fur-covered silhouette—two glowing yellow eyes of a mountain lion.

"Fuck... this might be it," I thought.

Adrenaline surged through my body. My pulse pounded in my ears. I dropped my pack, my hands moving quickly—hunting knife drawn, stick poised. The lion didn't budge. It stood there, muscles tense, eyes fixed on me, no more than fifteen yards away.

"Man… this would be one hell of a way to go," I thought to myself.

But I told myself that I would not go quietly. I was not ready to die. I made myself ready to kill again.

I threw my head back and yelled — loud, guttural, and primal. The lion flinched but stood its ground. I continued shouting, my voice slicing through the night air. For some reason, a wooden pallet lay discarded near the tracks. I grabbed it, dragging it in front of me and propped it against the power pole — a makeshift barricade. The lion remained still and stared at me, probably running its own calculus of whether I would be an easy kill.

I was ready to face this new, wild enemy of claws and fur.

My heartbeat began to slow, as I told myself this was just another fight. I decided not to wait for an attack but to launch one of my own. I threw a rock, and it landed near the lion's paws. No reaction. I threw another, this time striking the lion's body. It flinched slightly and shifted its position a bit. The stare-down continued — seconds stretched into minutes.

I kept yelling and throwing rocks, and the lion retreated a few paces.

I convinced the lion that I wouldn't be an easy kill, but maybe I wanted the fight. A deep, foolish part of me wondered what it would feel like — to kill the lion, to survive, and to tell that story. I kept shouting and daring the lion to attack me.

My vision became sharp in the night. I violently kicked the rocks around me and loudly chanted guttural noises like some paean of a Neolithic phalanx of warriors portrayed in cave art.

I was ready, but there would be no battle this night. I made eye contact with the beast and felt strangely connected to it. For a few moments, it was as if our souls touched.

The beast turned its gaze away from mine. Slowly and deliberately, it vanished into the shadows of the forest. I waited, listening and watching. I knew that predators sometimes circle back. I remained there, wedged between my pallet and the post, straining to hear movement in the trees.

The shadows of the forest played tricks on my eyes. But I saw nothing more of the beast.

Twenty minutes passed. Nothing.

I picked up the pallet—*just in case*—and carried it with me for perhaps a half mile. The night remained quiet; I finally dropped it and kept moving. I paused every few minutes like I did at war, but I neither heard nor saw anything.

I reached my truck around 10:00 p.m., muscles aching, mind still racing. And as I drove home, I found myself almost wishing I had killed that lion. I shook my head and exhaled.

I wonder now what became of it.

REINTEGRATION TRAINING

The Army recognized the challenges of transitioning from war to peace and provided a wealth of resources to support our return from war. For a couple of weeks, my unit hosted us at a hotel conference center and ran a series of reintegration workshops covering everything from finances to divorce. It also ensured we had access to psychologists and chaplains. Our leaders in the battalion went to great lengths to look after our physical and mental health as they constantly reminded us to speak to someone if we needed help.

We were permitted to wear civilian clothes to ensure comfort, and the training was insightful. The Army dedicated significant resources, undoubtedly spending thousands of dollars to bring in experts—therapists, counselors, and specialists trained in all aspects of reintegration.

I tried my best to stay focused, but I sometimes succumbed to daydreaming, staring at the clock, counting the days until I could leave. I wasn't searching for another lecture. I was seeking something more profound.

For some reason, I knew Hadrian's Wall was where I needed to be. I couldn't explain why, but I felt it deep in my bones. That walk — that journey — was what would put me back together.

I could find what I was looking for there. Maybe there, I would finally find peace.

WASHING OFF THE WAR

Historical martial cultures realized the challenges of reintegrating their warriors back into society. In some Native American cultures, warriors returning from campaigns were required to purify themselves before returning home. These purification rituals were conducted to cleanse them from the terrors of combat. They aided them in transitioning from their roles as hunters of men to return to peaceful village life. [1]

These rituals varied among different tribes. Among the Chickasaw, the war party leader sent a runner ahead, traveling a day's march in advance to notify the tribe of their return. In preparation, the tribe readied the war leader's winter house, a dwelling set apart from the rest of the village. The following day, warriors painted red and black approached the winter house carrying their enemies' scalps.

Upon arrival, they entered three days and nights of isolation, fasting from dawn to dusk. Throughout the day, they stepped outside to walk around a red-painted pole from which the scalps hung — a solemn ritual marking their transition from war back to peace. [2] After

---

1. Editors of Time Life, *The Way of the Warrior: The American Indians* (Alexandria: Time-Life Books, 1993), 102-4.
2. Ibid.

three days of isolation, the war chief, the warriors, and their female relatives participated in a solemn procession through the village, honoring families who had lost loved ones in battle. As they passed each home of the fallen, they tied a small piece of scalp to the house, a symbolic gesture of remembrance and tribute. Following the procession, the war leader and his assistant returned to isolation for another three days, while the other warriors were finally allowed to rejoin their people.[3]

The Maricopa of Arizona had a significantly longer and more rigorous purification process that began on the march home. Warriors who had killed or scalped an enemy were required to travel in a single file and in silence, following the main body at a distance. They also camped separately, drinking and eating very little. Upon reaching the borders of their homeland, they drank large amounts of water, after which a veteran would induce vomiting by inserting a plant stalk down their throats, symbolically expelling the impurities of war.

This marked the beginning of a sixteen-day period of isolation. Each warrior lived alone in a separate hut, bathed before dawn, and ate only one meal daily. For twelve nights, their heads were painted with mud and bark, demonstrating their purification. During this time, older warriors from the community visited them, offering guidance and reminding them of their responsibilities within the tribe.

Warriors who successfully took an enemy scalp were required to keep it with them for the entire isolation period. At the end of the sixteen days, they presented the scalp to the "Keeper of the Scalps," who stored it in a large ceremonial jar within the tribal meeting house. Before dawn on the sixteenth day, the warrior was finally allowed to leave isolation and return home. However, his reintegration was gradual. Upon entering his hut at sunrise, he sat at the rear of the dwelling, facing away from his family. Though permitted to eat with them, he

---

3. Ibid., 104.

could not move from his seated position except for sleep and bathing. This final stage lasted four additional days, ensuring his complete transition from warrior back to tribe member.[4]

The transition from warrior to civilian has never been easy. Across time and cultures, societies have recognized that those who have taken life and faced death cannot simply walk home unchanged. These Native American purification rituals demonstrated ways to cleanse the warrior's soul, allowing them to shed the burdens of battle before rejoining their communities. But in modern warfare, there is no red-painted pole to circle, no ceremonial fasting, no ritualized acknowledgment of what we've done or lost. Instead, we are given briefings, medical examinations, and access to spiritual and medical care.

But perhaps more importantly, warriors have always needed time, space, and purpose to find their way back. I didn't have a sacred ritual to mark my return—but I had the Wall. In the same way the Chickasaw warriors walked their path of purification, I knew I had to walk mine. Hadrian's Wall became my pilgrimage, transition, and reckoning with the past. I thought this journey could be a way to strip away the noise of war and find out who I was after the fight.

LEAVE

After weeks of medical screenings and reintegration classes, my battalion formed up on a lush green field at Fort Carson. We stood in parade formation—companies abreast, ranks aligned. The historian in me couldn't help but smile. I imagined us not as soldiers preparing for leave, but as warriors of another time lined up for battle in the nineteenth century. The rhythmic thud of boots marching across the grass filled the air, a sound I had always loved.

---

4. Ibid., 107.

At the command of "Fall out!" we moved to encircle our battalion commander in the familiar horseshoe formation—the Army's preferred way of delivering informal addresses. He gave us one final safety brief, reminding us to stay out of trouble, and wished us well on our long-awaited leave.

I exhaled. It was over.

I walked away, glancing back at the formation—at the soldiers and men of my old platoon. It was strange seeing us like this again, just regular people about to go on vacation. But the truth was, we weren't regular people. Not anymore.

We had just returned from hunting the most dangerous prey—man.

I was no longer a boy. I was a man in the truest, most unsettling sense of the word. I had hunted my own kind and had faced fear in its rawest form—the type that coils deep in the throat, grips the chest, and twists in the gut. And yet, I had done more than just endure it.

I had hunted fear itself.

And once you have hunted man, you are changed forever. I was ready to start my journey to Hadrian's Wall.

CHAPTER FOUR

THE WALL

LIMITS OF EMPIRE

I have been fascinated by the Roman Empire since I was a boy. I became hooked after watching classic films like the 1960 version of *Spartacus*. I vividly remember the scene where the legions deploy in perfect formation against Spartacus's army—something about their discipline, order, and sheer power captivated me.

I first heard about Hadrian's Wall in high school. I remember that lesson well. Every school has that one history teacher—the kind who reminds you of the instructor from *Starship Troopers*, passionate and intense. We covered the Roman Empire that day, and I was likely the happiest kid in the room. I was particularly fascinated by the legions. My teacher showed us battle maps, including the 52 BCE Battle of Alesia, and then projected the grand map of the Empire—a sea of deep imperial red stretching across Western Europe, North Africa, and the Middle East.

But something was off. As my eyes moved across Judea, Italy, Gaul, and finally Britannia, I noticed a problem—not all of Britain was red. A border cut straight across the island, leaving Scotland beyond Rome's grasp. This had to be a mistake. I raised my hand and asked why this was.

My teacher explained that the mountains and the fierce tribes of

Scotland were too much for the Romans. Instead of continuing their conquest, they built a massive, fortified wall across the island. And then, for the first time, I heard the name: Hadrian's Wall. This was Rome's frontier, the very edge of the empire, the line they could not cross.

I spent the rest of the lesson staring at that map, lost in thought. How could the most disciplined war machine in history—an army that had conquered much of the known world—be stopped? What made these northern tribes so formidable? I had just seen the then-newly released *Braveheart* and imagined fierce warriors in woad paint clashing with my boyhood heroes—the legions.

But in high school, more pressing matters—girls, football, video games—distracted me from solving what seemed to me to be Rome's great mystery.

Years later, after my initial enlistment in the Marines, when I returned home to Michigan intending to become a high school history teacher, I revisited this question. In 2002, I enrolled at Eastern Michigan University, eager to pursue a degree in history. By my second semester in 2003, I decided to finally pursue the question that had haunted me since high school: "Why couldn't Rome conquer all of Britain?"

I planned to write a research paper on it. But before I could begin, war came for me instead, and I was called up for the impending war in Iraq. Yet, the question of Hadrian's Wall never left me. Over the years, it remained lodged in my mind, resurfacing again when I went to graduate school in 2005, where I shifted my focus to Julius Caesar's 58-51 BCE Gallic War. There, I continued my studies on Roman military strategy and became even more fascinated by the sheer combat power of the legions. I studied battles like Cynoscephalae, Zama, and my favorite, Alesia, marveling at Rome's ability to crush its enemies through discipline and tactical brilliance.

And yet, my thoughts kept drifting back to that map—to that

border in England. How had these invincible legions been halted at the foot of the highlands?

Only later did I come to truly understand the deadly combination of rugged terrain and an unorthodox, lightly equipped enemy. These tribes didn't fight like the Gauls or the Carthaginians, gambling on massive, pitched battles in open fields. Instead, they fought like ghosts, striking and vanishing into the mountains they intimately knew.

Rome did attempt to conquer Caledonia (modern-day Scotland), bringing the tribes to battle at Mons Graupius in 83 AD. They claimed victory, but the land was never truly subdued. The legions could win battles, but they could not hold the highlands.

So, instead of continuing their war, Rome built a wall. A wall that marked the limit of their empire. A wall that had called to me for years.

In 2009-10, I faced my own challenges in taming the highlands, which led me to uncover the reason behind Rome's failure to conquer these northern lands. The answer wasn't buried in the writings of Tacitus, Polybius, or even my beloved Caesar. It wasn't hidden in archives or revealed through archaeological digs.

Instead, I found my answer in the most unexpected of lecture halls. I discovered my answer in war. While fighting in the rugged mountains of Afghanistan, I came to understand what no book could ever fully convey: the harsh reality of waging war against an enemy who controls the terrain. In the highlands, I confronted a fierce, unconventional, and determined enemy that knew every pass, valley, and hiding place. The mountains weren't just their home; they were their greatest weapon.

Only then did I truly grasp what had stopped the Roman legions at the edge of Caledonia. I believe I captured the sheer difficulty of mountain warfare in the following passage from my 2012 book, *Siren's Song: The Allure of War.*

US AND THEM

We wear and move in heavy armor. Our weapons and munitions cost thousands of dollars. We can see in the dark, seeing humans in the form of heat signatures many kilometers away under the cover of darkness. We can call for an array of air support, ranging from unmanned aerial vehicles, F-16s, A-10s, to Apaches, and Kiowas. We can drop 2,000-pound laser guided bombs. Our 155's can hit targets, accurately, from miles away using GPS. We can MEDEVAC the wounded and save life. We can air assault supplies and ammo within minutes of a request. We have satellite imagery and can communicate instantly with our comrades utilizing state of the art communication equipment. We are loud, menacing, and, most of all we are ungodly powerful.

Our opponents wear no armor. They wear linen and carry just a few magazines or a few RPG rounds. They have no laser sights on their weapons, any air support, or satellite imagery. They do not have platoons of men waiting to be a Quick Reaction Force. They can't call for fire on the move, nor see in the dark. They cannot MEDEVAC their wounded. However, they have the heart to persevere. They fight against overwhelming odds and do amazingly well.

This Afghanistan, with her mountains of stone and heat, hides the fighters within her folds of stone and jagged rocks. These guerillas move like shadows in the forest and attack from relative safety. The mountains are their armor; they serve as their air support, and as their reconnaissance platforms. The mountains are theirs. They can blend into the rocks as well as the populace that surrounds us.

Even blessed with the advantages of overwhelming firepower, I learned perhaps one of the few truths that have stood the test of time throughout military history: "Don't mess with mountain people." There are several reasons not to. First, mountain people possess mastery of the rugged terrain as they intimately know every valley, cliff, pass, and hiding spot. Their familiarity with the land allows them to utilize natural fortifications and maneuverability to hinder invaders unfamiliar with mountain warfare.

The mountains provide ideal locations for ambushes and guerrilla tactics, enabling numerically inferior local fighters to inflict casualties on better-equipped but less-adapted opponents. Second, living in mountainous regions builds mental and physical endurance. Locals grow up walking on rugged trails, enduring harsh weather conditions, and surviving with limited resources, naturally toughening their bodies and minds. Third, operating in mountainous areas presents enormous logistical challenges for invaders, straining supply lines and mobility. Armies that rely heavily on complex logistical support, like tanks, vehicles, or in Rome's case—wagons—find themselves vulnerable and quickly bogged down. Additionally, mountain people often exhibit strong cultural cohesion within tribal structures, making them very wary of outsiders.

Perhaps some places on Earth are better left marked on the map as "Here Be Dragons." I carried the frustrations of the legions with me on my epic journey to the Pech River valley in Kunar, Afghanistan. Mountain people rarely give up easily. As I trudged through the valleys of Afghanistan, I realized that the Romans must have experienced the same frustrations as we did. Despite their unmatched discipline and organization, they could not conquer Scotland. Like the tribes of Caledonia, the guerillas in Afghanistan fought on their own terms, using the terrain to level the playing field against our technology.

I realized Hadrian's Wall was not just a frontier, but an admission. It marked the limits of an empire that knew it could stretch no further, a line drawn, not only against the tribes of Scotland, but also against the very nature of the land itself. This was a lesson I carried with me through my battles — a reminder that even the greatest empires have limits defined not just by the strength of their enemies, but also by the will of the earth itself to resist domination.

BUILDING THE WALL

After visiting England, Emperor Hadrian ordered the construction of the wall to begin in 122 AD. It extended over 73 miles across England, from Bowness-on-Solway in the west to Wallsend in the east. No other Roman frontier fortification matches the scale of Hadrian's Wall. Today, remnants of the wall remain as a testament to the Empire's limits. It is now a World Heritage Site, and visitors can walk the trail all the way from Bowness-on-Solway to Wallsend.

There are no surviving Roman documents that explicitly state why Hadrian's Wall was built. The only direct mention in ancient texts comes from the 4th-century *Historia Augusta*, which records: *"Hadrian was the first to build a wall, 80 Roman miles long, to separate the Romans from the barbarians."* Another inscription near the wall states: *"After the barbarians had been dispersed and the province of Britain had been recovered, he added a frontier line between either shores of the Ocean for 80 miles."* [5]

Aside from these brief mentions, we're left to speculate. Some scholars contend that the wall was mainly constructed for economic reasons, controlling the flow of goods and people across the border. However, I firmly disagree that that was the only reason.

I believe the wall was constructed in response to the nature of guerrilla warfare in this untamed frontier of the Roman Empire. While Roman sources dutifully recorded significant military cam-

---

5. *Historia Augusta*, trans. David Magie (Cambridge, MA: Harvard University Press, 1921).

paigns, more minor skirmishes fought in the shadows of the high-lands and the forests beyond the frontier were unlikely to be recorded. They would have been too infrequent, scattered, and insignificant in the grand scale of empire-building to merit detailed records.

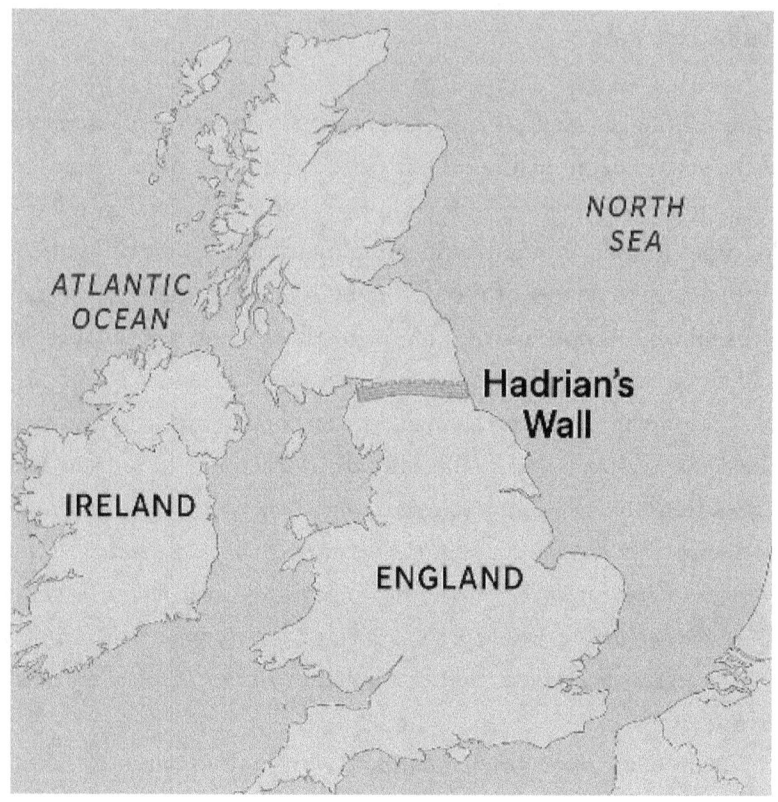

Hadrian's wall.

The American experience in the Global War on Terror provides a modern parallel. While a handful of battles — COP Keating, Wa-nat, Sadr City, Fallujah — have been immortalized in books and films, they represent only a fraction of the war. The vast majority of en-gagements — thousands of firefights, millions of rounds expended, countless RPGs, and IED blasts, will never be remembered. Some will merit a few lines in official histories, while others will exist only in the memories of the men who fought them or in the few memoirs that come into existence.

I believe something similar occurred on Rome's northern frontier. There must have been dozens, if not hundreds, of minor skirmishes—ambushes, raids, and sudden clashes in the misty hills—before Hadrian concluded that a physical barrier was necessary. Some recently discovered reports on Roman troop strength even reveal injuries sustained in combat-related incidents along the frontier, further supporting the idea of ongoing, small-scale conflict.[6]

The Romans could only endure so much. Just as I witnessed in Afghanistan, there is a limit to what soldiers, steel, and firepower can achieve against an enemy whose fortresses are the mountains themselves. Even in the 21st century, advanced armies struggle to hold territory against an enemy that knows the terrain better, moves faster, and fights on their own terms. The Romans must have learned the same harsh lesson in Caledonia—and Hadrian's Wall was their answer.

MY DECISION TO WALK THE WALL

During May of 2010, in my final month of deployment in Afghanistan, I prepared myself for my voyage home. But home, as I once knew it, no longer existed. Instead of returning to peace, I would be stepping into the firestorm of a dissolving marriage. On top of that, I wasn't sure how much longer I wanted to stay in the Army.

I had experienced combat, and with that came a question I couldn't shake: Had I completed my journey as a soldier? I needed something, an escape, a break, an adventure to help me find clarity. More than anything, I needed peace.

---

6. Adrian Goldsworthy, *Hadrian's Wall* (New York: Basic Books, 2018),10.

One early morning on FOB Blessing, after a workout, I sat down to breakfast, flipping through my favorite book—Caesar's account of the Gallic War. I carried this book with me on every deployment or field exercise. That morning, I was reading about Caesar's first invasion of Britain in 55 BCE. I traced my fingers across the map of Britannia, my eyes drifting north—to the rugged mountains.

And then it hit me.

The answer to a question I had first asked years ago—Why couldn't Rome conquer all of Britain?—was staring right back at me: *the mountains.*

I glanced up at the jagged peaks of the Pech Valley, standing like ancient sentinels guarding the land. I thought about my war—the endless battles in these highlands, the way the enemy would always slip away into the mountains, using them as both a shield and a weapon. No matter how many firefights we won, the mountains always won in the end. I was sure that the legions of Rome must have faced the same fate.

I thought of Hadrian's Wall. Maybe, like us, the Romans had realized they would never fully conquer the highlands and the tribal peoples of that region, so instead, they built a barrier to hold the line. Perhaps we should have done the same. At that moment, I knew I had to travel there.

I immediately went to FOB Blessing's recreation building and found a computer to research the feasibility of walking Hadrian's Wall. To my surprise, it was a popular trek in the UK. I dove into the details—where to fly in, which trains to take, and how long the journey would be. And without hesitation, I booked my ticket from Denver to the United Kingdom to depart in the final week of June 2010.

I also decided I didn't want to do this alone. I invited two of my closest friends, Isaac and Seth Williamson, to join me. We had met years earlier at Eastern Michigan University, where we had lived together in Ypsilanti, Michigan. I had run marathons with both of them, so I knew they had the endurance for the long trek.

Just the thought of the Wall—its history, its meaning—already felt like medicine for my wounded heart. I had yet to set foot on it, yet somehow, its very existence was already pulling me forward, giving me something I hadn't felt in a long time: a sense of purpose. I couldn't wait to walk it

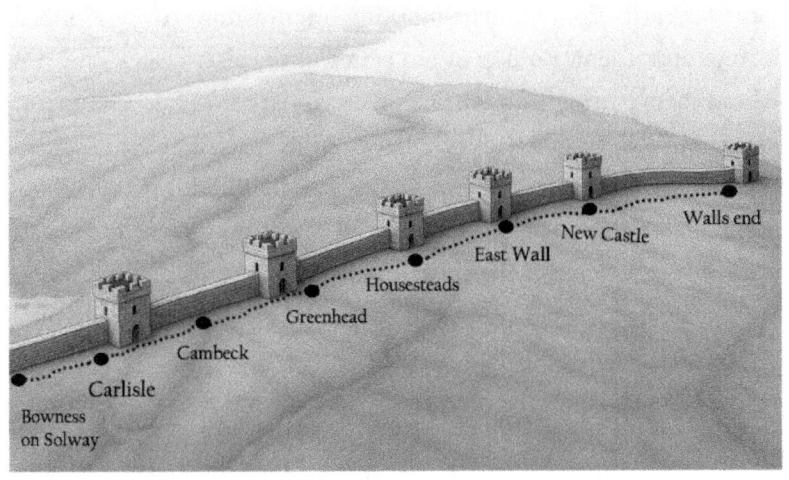

My major stops along the Hadrian's Wall trail.

CHAPTER FIVE

DAY 1, JUNE 21, 2010: THE WALK BEGINS

I awoke early that Sunday morning in my downtown Colorado Springs apartment. Rolling over, I stared at the alarm clock, its bright red numbers cutting through the predawn darkness. For a moment, I lingered in that fragile stillness, that fleeting space between sleep and wakefulness where time seems to hesitate. In these rare, weightless moments, we choose to rise and face the day or steal a few more heartbeats of refuge beneath the covers.

For the past year at war, stillness had been a luxury I could rarely afford.

In Afghanistan, some mornings were dictated not by soft beeping alarms or the slow bloom of daylight but by the raw violence of RPG impacts—thunderous, concussive blasts that rattle bone and earth alike. These wake-up calls were not gentle—they came as the sharp, percussive cracks of incoming gunfire, ripping through the fragile veil of sleep and hurling me into another day of survival.

I remember them still.

And even today, thousands of miles away, in a world wrapped in peace, I can still hear them.

MAY 2009: COP MICHIGAN, PECH RIVER VALLEY, KUNAR PROVINCE AFGHANISTAN

POP, POP, POP! POP- POP, POP-POP, POP-POP! BOOM-

BOOM, BOOM-BOOM, BOOM- BOOM-BOOM! BA- BAM, BA- BAM, KA-KA! KA- KA! KA- KA!

Gunfire and RPG explosions are the most unforgiving of alarm clocks—they don't come with a snooze button.

My eyes snap open. I am awake—the most awake I have ever been. The early morning sun leaked through the gaps in the doors of what passes for our barracks. This building looked more like a Boy Scout cabin than a soldier's quarters.

The gunfire continues, a terrible rhythm—the foreboding song of death. Time slows to a crawl, the way it always does in battle, where seconds stretch into eternity, and fights play in our minds for years. I have heard gunfire my whole life—on the rifle range and in the Michigan woods.

But gunfire sounds different when it is meant to kill you.

I blink, clearing my vision, and watch dust particles drift carelessly in the sunbeams. In that moment, I envy them—their weightless dance, oblivious to the chaos around them. No fear, no urgency, just an effortless ballet in the light. I cling to those last heartbeats of peace, savoring them before the adrenaline kicks in—venom flooding my veins.

Then I hear him.

"Salinas! Get the hell up! Get your fucking kit on!"

Lt. Kligensmith stands over me, bare shoulders under his body armor, wearing only PT shorts, running shoes, and a helmet with the chin strap unbuckled. He looks ready for the beach, not a firefight. But the look in his eyes gives me pause. It isn't fear.

It is acceptance.

A quiet understanding that Death has arrived. That this is happening. And it is time to dance. Lt. Kligensmith had been in Kunar for nearly a year as a part of the 1st Infantry Division,1-26 Infantry that we were relieving. He stares at me, and in a single heartbeat, his eyes tell me everything a warrior dares not speak:

"We are about to run straight into that storm of bullets. This might be our last breath, so take it in. You may be moments away from hot lead slicing through your body, tearing your intestines into the dirt. A round might snap your spine, paralyze you, drown you in your own blood. But don't worry—we'll do the same to them. We are better trained. We are stronger. We are lions, and we will rip their throats out. Follow me, and let's send these fuckers to their graves."

I don't hesitate. I pull my body armor over my sweat-drenched skin, strap my helmet tight, and grab my rifle.

"Okay," Kligensmith says. "We're sprinting to those HESCOs across the courtyard as soon as I open the door. Ready?"

"Hell yeah," I breathe.

"Follow me!"

The door flies open, and the furnace of Afghan heat hits. The sky itself screams as the air erupts with terror. Rounds snap past. Dust kicks up. The world narrows to a single, brutal purpose—survive.

We run.

We slam into the first line of HESCOs, and my first battle begins.

* * *

Back in my apartment, I sat up slowly, shaking off the last remnants of sleep. Across the room, my hiking clothes and boots rested, folded neatly on a chair, waiting for me like old friends. A familiar excitement stirred in my chest. I had chosen my gear with care—"contractor casual"—a tan long-sleeve shirt, combat trousers, and the same hiking boots that were issued to me in Afghanistan.

I ran through my final checks, ensuring everything was in place: extra socks, foot powder, pants and shirt, moleskin, bandages, a map, Advil, a CamelBak, sleeping bag, a tent, a notebook, running shoes, compass, and some granola bars. I checked for my passport—again and again—twelve times. I didn't shave. I decided to let my beard grow, a quiet tribute to the bearded men I had once hunted. If I was walking Rome's border with their so called "barbarians," it only felt right.

Stepping onto the balcony, I cradled my last cup of coffee in Colorado Springs. The Rockies loomed to the west, their dark silhouettes sharpening in the twilight. My eyes traced their jagged edges, and I thought of the men who had taken my place in Kunar. Somewhere in those mountains, Americans were fighting for their lives.

I took a slow sip, savoring the taste, savoring the simple act of tasting. Some men would not be so lucky.

STEVEN DREES

He was the first person in our company to die. He was shot in my Area of Operations. A single shot hit him while he was in the gun turret. He was killed only 1.5 miles or so from where I slept. I discovered it while on a resupply mission at Asadabad. I spoke to some men in his platoon, such as David Klutenkamper. In their eyes, I saw the distinctive sadness of a warrior for the first time — it was like watching lions tear up.

He was one among many who were too young to die. I still see his face, so young that I never saw stubble on it.

I remember him not at war, but at peace. It was one of the last few weeks before deployment, and there wasn't much left to do besides show up to base and conduct PT.

Most of the company was in the gym, doing a light lift. I saw him spotting one of his friends. He smiled at me and nodded. I smiled back.

"Making gains, Drees?" I asked.

"Yes sir," he replied

When I heard about his death, I saw his face again. A clean, young, innocent face with innocent brown eyes.

Within a week or so of his death, we held a memorial at the near-by COP Michigan. The Company's First Sergeant took his post for his part in the service and began his roll call.

"Pvt Smith!"

"Here, First Sergeant!"

"Pvt Johnson!"

"Here, First Sergeant!"

"Pvt Drees...." This time, there was no reply.

"PVT Steven Drees!"

"PVT Steven Thomas Drees."

The air stung with silence. A slight breeze danced through our ranks, cooling us somewhat under the hot Kunar summer sun. Behind the formation, I glanced at my men standing shoulder to shoulder in their dirty uniforms. Their shoulders looked so strong and impressive in their gear. I wondered if I would soon have to attend a memorial for one of them. This fear is the water that can split the rock of a warrior's heart.

"COMPANY! ATTENTION!"

"PRESENT... ARMS!"

The national anthem played loudly, booming to the corners of our small frontier fort and echoing off the valley. Chills ran down my spine in pride as our nation's hymn played here. Typically, the anthem precedes sporting events and other cheerful ceremonies. But here, Death claimed it and made it her song.

Steven Drees was the first of our company to fall. But he was not the last.

Tears rolled from my eyes as I realized the anthem had been transformed. It was no longer a song of honor to our nation. Instead, it had become the lullaby that put our dear comrade to sleep. He would never again rise.

"Order... arms. Parade... rest."

The bagpipes played "Amazing Grace." It was the warrior's ode. A warrior's instrument paying homage to a fallen warrior.

The company fell out to pay its final respects.

The line shuffled me closer to his memorial. A rustic wooden crucifix held up his boots, rifle, helmet, dog tags, and picture—a proper monument to a man at arms.

I approached the memorial and gave Pvt Drees one last salute. I stared into the face of this warrior, not a man with a weathered beard and wizened face that had fathered children, but a smooth and innocent countenance. I knew him well enough; he could go a few days without shaving before anyone would notice. I saluted, and my eyes fixed on his pre-deployment photo, which had been taken for this very purpose. I became lost in his eyes as if his soul inhabited the likeness. There were so many things I wanted to say to him.

"I am sad that you are only nineteen and that you died here in this god-forsaken valley. You should not have been here. Instead, you should have been drinking beer on a college campus. I want you to know there is more to life than dying with honor. I wish you could have known that before coming here.

"I wish you could have met the girl of your dreams. I wish you could have felt your child kick while still in your wife's womb. I wish you could have seen your child grow and call you 'daddy.'

"Instead, your life was stolen by an incoming bullet, and you fell from your gunner's turret."

The ceremony concluded, and the Dagger Platoons mounted up and returned to their respective outposts. I drove east, praying for a fight. Yet, the land seemed quiet.

However, as we neared home, my prayer was answered. I saw plumes of smoke rising from our outpost as we drew near. COP Honaker Miracle was under attack.

My platoon drove into the COP and assumed battle positions as the firefight raged on.

* * *

LEAVING FOR THE WALL

I slung my heavy rucksack over my shoulders and stepped outside. The cool predawn air of a Colorado summer enveloped me, carrying the soft hum of distant cars. I strolled, my pack filled with supplies for peace, not war. No 5.56, no linked 7.62 ammo, no night vision goggles, no body armor. Just me, my notebook, and my journey.

I felt like a stranger in this world of steel, brick, and paved roads. Civilization no longer felt like home. My world had been dust and mountains, outposts and ambushes. I had spent a year in a landscape that resembled something from the Bible — but with machine guns. I missed its raw simplicity. I missed Afghanistan.

I missed my war.

I listened to the rhythmic patter of my hiking boots and the gentle creak of my pack. This journey was meant to take me to Hadrian's Wall, but I would later realize that the hike never truly ends. It is a journey that all warriors undertake. The thing about coming home from war is that you return not only to an uncomfortably unfamiliar place, but also to a version of yourself you no longer recognize. You are no longer an instrument of war, a harvester of men. You are expected to be civilized again, to adjust to a world that hasn't changed while you have become something else. Perhaps this is why we struggle — because we are not just strangers to the world.

We are also strangers to ourselves.

I walked through the empty streets of downtown, past shuttered bars and silent storefronts. Reaching the overpass above I-25, I stopped, watching the quiet convoy of early morning commuters speeding beneath me. Order. Efficiency. Motion. I thought of the roads I had left behind. Roads where nothing moved fast. Roads of dirt and blood. Roads that could break the axle of an MRAP, flip a 14-ton hull, or turn a simple patrol into a day-long firefight.

QATAR KALA FIGHT, JUNE 2009

The morning was just warming up into another hot, Afghan summer day. We held our positions in the village, fighting what we thought would be a skirmish. We took cover by a mud-brick wall and dueled with the ghosts in the mountains, who only temporarily appeared to us in the short glimpses of muzzle flashes or linen as they bounded to and from pieces of cover.

"Okay, get ready to move out. We are heading out of here in five minutes. Salinas, get back here," the outgoing commander said over the radio.

Sgt. Moffet and I held the right flank of our dismounted unit. It was time for us to return to the company headquarters. We stood at the corner of a building, which appeared to be a small barn. I hesitated before giving the order and felt envious of the two cows who casually munched on their hay while foolish men continued their heated argument.

I looked at Moffet; it was time to go. We had to bound back one at a time from building to building. The enemy knew where we were and was trying to hit us with bursts of machine gun fire.

I sprinted across a large expanse of maybe fifteen meters and called for Moffett to move.

"Okay, go!" I called to him.

As he was about to run out, the alleyway erupted in a mass of bullet impacts.

Moffet jumped back, paused, then ran out again.

Soon, we returned to our company HQ and looked forward to pulling out. "Roger, wrecker ETA four hours," I heard my commander say. We would end up fighting it out for another six hours until the disabled MRAP could be recovered.

* * *

I arrived at the Holiday Inn, where I would catch the shuttle to the Denver Airport. Sinking into one of the cushioned chairs in the lobby, I pulled out my maps of Hadrian's Wall and traced the route with my fingers. The contrast of my new environment felt jarring—this polished, air-conditioned world of civilians carried on as if nothing had changed, while I felt like a ghost wandering unseen through a life that no longer fit. Colorado had not changed, but I had.

I waited, restless and eager to leave. Maybe I wouldn't feel so out of place abroad, in a foreign land. Maybe being a stranger somewhere else would feel more normal than being a stranger at home.

The shuttle finally arrived, and I climbed aboard, stowing my gear in the back before settling in for the hour-long ride. A few other travelers joined me.

"Are you going on a private security contract?" a young woman asked, studying my boots and rucksack.

I chuckled. "No, just a hiking trip. My little gift to myself after deployment," I said.

"Oh… well, you look like one of those guys," she said.

I smiled. "I get that a lot," I said.

She was eager but nervous about heading to ROTC summer camp. I gave her some advice, having experienced it just four years earlier. Another passenger, a young US Army private, was deploying to Afghanistan in a few months. I told him to keep his head on a swivel and gave him the measured encouragement veterans often provide to the next wave.

I have always liked airports, crossroads of travelers, each with their reasons for running toward or away from something. That's life in the twenty-first century. We rush and rarely stop to take stock of things. But today, I was done talking. I wasn't accustomed to long, casual conversations anymore. My world had become one of brevity—radio calls, short commands, and clipped reports. I leaned back, watching the Rockies roll past through the window, just as beautiful as they had been the first time I saw them, yet now something foreboding as well—fortresses, barriers, the domain of the enemy.

The first time I saw the mountains of Afghanistan, I thought, *"So, this is why the Soviets couldn't win."* I had studied military history, but I never truly grasped the complexities of mountain warfare until I experienced it firsthand. We could enter the mountains and wage battle within them, but we could never truly own them. Our bombs, artillery, and bullets could briefly occupy those peaks, but our lease was always short and always paid in blood. The valleys—Pech, Watapor, Waygul, Korengal—held our stories, victories, and losses. They silently bore witness to the cost of war.

As I stared at the Rockies, I thought of Rome's legions struggling in the Scottish highlands, ultimately choosing to build a wall rather than claim the land. Maybe some places aren't meant to be conquered.

MOUNTAIN AMBUSH, JULY 2010

BOOM! Crack, crack, crack! RPGs and gunfire echoed off the mountain. As the road erupted with the impact of enemy munitions, it felt as if the mountain itself was trying to kill us, as if it had become alive and breathed fire into our ranks, roaring with tracers and explosions. And yes—that Dragon roared. The roar of incoming fire. How does one describe this? How does one convey the sound of death?

Before my first fight, I thought I would be brave. We all do. But the first time I heard the roar of an RPG; it shattered any preconception I had of bravery. Instead of a hunger for glory, I felt different things:

A chill in my blood

A 200-pound plate on my stomach

A dumbbell on my throat

My hand coated in sweat

All I want at that moment is life

My body screams that I want to live

I want the fire to stop.

The animal brain demands action: Fight or flight?

We fought not just because it was our job. We fought to keep breathing and keep our hearts pumping blood. Against this Death, we stood. We roared back. We had to kill them before they killed us. We had to shoot them and break their bodies.

We had to burn the mountains, melt the rocks, and destroy the men within them. We had no choice. I wanted to breathe and survive. I tried to keep my men alive.

My men pounded back. I gazed at the mountain and saw the enemy. I established map coordinates for artillery. I desperately asked the war gods to fix their cannons on these grids. I whispered.

From a whisper, I asked for death

From a whisper, I called for thunder.

I heard our artillery rounds come in. Nothing else in the world sounds like an artillery round falling through the air. The 155mm rounds tore the air as they fell. They seemed to scream: "Here I am. Fucking run. I am here to end you."

From a whisper, I commanded bullets. Our .50 caliber machine guns with their distinctive "KA KA KA KA" pounded the highlands to the sound of cascading expelled brass falling from our guns. We moved closer to the dragon. Together, we moved toward death.

* * *

I will never view mountains the same way again. Even now, fifteen years after my war, whenever a hill rises on the horizon, a sharp, familiar sliver of concern embeds itself deep within my soul—a silent reminder. I still love hiking mountains, but I know what they can do. Mountains are more than just landscapes; to me, they are fortresses and battlefields.

Certain truths accompany those of us who have hunted men. War does not fade; it burrows into the heart, mind, and soul, becoming part of you forever. Even nature itself changes. I see it in the mountains, but I also see it in the fields. Cornfields — an innocent sight to most — still haunt me. In the heat of the Afghan summer, those towering stalks became sanctuaries for the enemy, who used them for cover to ambush us. I was spared from an attack in them, but the possibility of that unseen threat still lingers, an unease that never quite fades.

After about an hour, our shuttle turned east, leaving the Rockies behind as we approached Denver International Airport. The mountains gradually blended into the plains, with vast expanses stretching endlessly from the peaks. This airport is unlike any other I've seen. Rising from the ground, its large white rooftops resemble massive teepees, ghostly against the sky — a strange final monument before departure. We arrived, and I quickly gathered my gear and made my way to check in for my flight to Manchester. I had another border to cross, another journey eastward.

I didn't fully understand what I was walking into, but I had to take that step. This was not a vacation. It was a journey into myself, my past, and the quiet echoes of war that still cling to me. It was the first leg of a long, slow, and unending march away from Death.

CHAPTER SIX

DAY 2, MONDAY, JUNE 22, 2010: BOWNESS-ON-SOLWAY TO CARLISLE

As a writer, I always strive to be productive on transatlantic flights, convincing myself that I will spend hours writing or catching up on reading. However, those hours are usually squandered on a mix of movies and sleep. On this journey, exhaustion took precedence. Knowing I would need to start my walk soon after landing, I surrendered to rest. Like many soldiers, I have honed the ability to sleep anywhere and at any time.

After a brief stop in Atlanta, I continued my journey to the United Kingdom. Sleep overtook me once more until the descent jolted me awake. As the plane began its approach, excitement surged through my veins. This was my first trip to the United Kingdom, a place I had long envisioned but had never visited.

I gazed out the window and caught my first glimpse of England—the rolling green hills stretching toward the horizon, wind farms rising like modern watchtowers against the sky. It was a serene sight, one that felt worlds apart from my previous flights into war zones. There was no roaring C-17, no combat gear weighing me down, and no looming danger waiting on the tarmac. Just a quiet descent into a land steeped in history, where my journey across England and into my warrior soul was about to begin.

LANDING IN BAGRAM, MAY 2009

Alright, helmets on… The plane descended at a steep angle, sending adrenaline rushing through my veins as my stomach lurched. I clenched my teeth, swallowing the familiar wave of nerves. Below, Bagram Airfield sprawled out beneath the Afghan sun, worlds apart from the one we had left behind. Even inside the aircraft, I could already feel the heat pressing in, thick and relentless, a prelude to the furnace waiting outside.

The tail of our giant bird of war opened, and the heat rolled in like a wave from a gigantic oven. I squeezed my rifle tightly, anticipating the unknown challenges to come. As we debarked the aircraft, I peered through the convection waves in the distance. The Hindu Kush Mountains, stone guardians, rose in the distance to greet us, warning of the dangers deep in their valleys.

Stranger, go away
You are at the end of the world
Go no further
Here is only death
Here is only pain
My rocks will tear you apart
My valleys swallow men, dreams, and armies
Many have come here
Many have fought here
But their flags have left

The Hindu Kush waited for me. Somewhere in those valleys, my fate was sealed. There, I would hunt. There I thought I would die.

Unexpectedly, I left those valleys. But they never left me. I just need to be alone to return to them. I can still smell them—sweat, mud, sand, oil, exhaust, and the sharp sting of gunpowder.

* * *

I stepped off my plane in Manchester, an ocean and a world away. My walking stick, considered too dangerous for the cabin, awaited me at baggage claim. The customs agent welcomed me with polite curiosity.

"Purpose of your visit?" she asked.

"I'm here to walk Hadrian's Wall," I said.

Her face lit up. "Really? I just saw two brothers who said they were doing the same thing."

"Would their last name be Williamson?" I asked.

She glanced at her screen, then smiled. "Actually… yes."

"I'm meeting them here—we're walking the wall together."

With a nod and a smile, she stamped my passport. "Well then, best of luck to you."

Beyond the checkpoint, my old friend Seth waited. Towering at 6'7", he was an imposing figure, yet I had always known him to be one of the kindest men I'd ever met. His brother, Isaac, was shorter but equally steadfast—a rock in his own way. I met them both in 2005 at Eastern Michigan University. We are good friends who lived together and laughed together, and I cherish their friendship to this day.

"Where's Isaac?" I asked.

"Downstairs with our stuff," Seth replied.

Isaac greeted me with a brotherly hug. I felt something close to normal for the first time in a long time. With them, for a moment, I could forget, pretend I was just another guy heading on an adventure, not a soldier coming home from war.

But as much as I tried to ignore it, I knew the truth—I was different now. Seth picked up on it. "How are you feeling? Any effects from the war?"

I forced a smirk. "I'm good. Officers don't get PTSD," I said, arrogance masking the truth.

We boarded the train north to Carlisle. As an American, train travel was still novel, a welcome change from the convoys and helicopters I had grown used to. I sat by the window, watching the countryside roll past, my body heavy with exhaustion, whether from the war or the jet lag, I wasn't sure.

The English landscape unfolded before me, a vivid patchwork of green fields and stone cottages, dotted with flocks of grazing sheep. It looked exactly as I had imagined, a scene from an old painting, untouched by time.

It had been years since I had seen green like this. With its high desert terrain, Colorado was golden and dry, beautiful in its own right, but nothing like this. Afghanistan had valleys that could bloom, but nothing compared to the rolling pastures of England.

For the first time in years, I stared at a landscape that wasn't trying to kill me.

And yet, even here, the mountains still haunted me.

I thought of the Pech, the Korengal, the valleys where we had fought and bled. I thought of Rome's legions, struggling through the highlands of Scotland. Perhaps some places were never meant to be conquered.

What was most refreshing about this train ride was that, for the first time in a long while, nothing reminded me of war. My mind felt light and unburdened as I gazed out at picturesque villages and sprawling fields rolling past my window. I felt at peace. I was thrilled to be alive, free from the grip of war. Silently, I thanked God for sparing my life, especially during those countless moments when a bullet came inches from tearing my flesh.

During the ride, I wandered into the dining car and savored my first taste of English chocolate. The rich sweetness melted on my tongue, transporting me to past memories of watching a trolley full of treats roll down the Hogwarts Express' corridor in *Harry Potter*. I felt an unfamiliar burst of pure, innocent joy—the kind I thought that war had stripped away.

As we continued north, the landscape opened into the breathtaking highlands of the Lake District, green and serene. Yet, as the train cut through narrow mountain passes, a familiar unease crept into my chest. Even in their peace, the highlands echoed something deeper, ancient and untamed, making me fear war hiding in wait behind every ridge. Even here, thousands of miles from Afghanistan, I couldn't shake the feeling. Mountains always have their own way of keeping secrets.

In the mountains, here be dragons. Here be RPGs. Here be PKMs. Here be DSHKs. Here be 107mm and mortar tubes.

We did not control the mountains as much as we pounded them.

The Romans knew as we did what mountains and mountain people bring.

Some places cannot be conquered.

Some places remain forever dark.

After two hours, the train began to slow as we neared the outskirts of Carlisle. My pulse quickened, anticipation swelling in my chest. We were approaching the starting point of Hadrian's Wall, and I could almost feel it calling to me, its presence growing stronger with each passing mile. It wasn't just the frontier of Rome I was nearing — it was the frontier of my soul. I was eager to step onto the trail, to lose myself in the path where history and hiking would merge.

When the train stopped, we made the brief, five-minute walk to the bus station, arriving just in time to catch the #93 bus to Bowness-on-Solway. We boarded shortly after 1 pm, weaving through the charming streets of Carlisle. The UK felt like a time capsule, with its old-world architecture, industrial-era brickwork, and narrow roads lined with cars that looked distinctly foreign to my American eyes.

As we left the city, the landscape unfolded into the serene beauty of the English countryside. Rolling fields stretched endlessly, divided by ancient-seeming stone walls, dotted with villages — small clusters of cottages and farmhouses which seemed to exist outside of time. The roads, barely wide enough for two vehicles to pass at once, twisted through this pastoral maze like a footpath. I marveled at how deftly the bus navigated such tight turns, its wheels hugging the edges of the narrow lanes.

The English countryside on the train ride north to Carlisle

Before long, we stepped off into the quiet village of Bowness-on-Solway. The air was crisp, carrying the scent of the nearby sea, and the stillness felt almost sacred. Ahead of us, the trailhead awaited—the start of my long-anticipated journey along Rome's northernmost border.

Trail marker in the village of Bowness-on-Solway.

BOWNESS-ON-SOLWAY

The village marked the western starting point of Hadrian's Wall and was once home to the Roman fort of Maia. Standing there, I couldn't help but draw parallels between these ancient outposts and the small combat outposts (COPs) I had lived in along the Pech River Valley. Like Maia, our COPs, sometimes no larger than 200 meters by 150 meters—were meant to be bastions of security, but often they only provided the illusion of it. In Kunar Province, some of our outposts had been overrun; places such as Ranch House, Wanat, and COP Keating still echoed in the annals of war as places where the enemy breached our perimeters.

Compared to the fortified structures of Rome, our defenses were crude, but the purpose remained the same: to hold the line against our enemies.

On many nights, I stared out into the highlands of Afghanistan, much like Roman sentries once gazed across this frontier. We, too, were the far-flung outposts of an empire, banners marking the limits of control. The Romans eventually accepted that some lands could not be tamed and built a wall to separate order from chaos. Our version of that wall was a chain of COPs, fragile islands of fire and steel scattered throughout the Afghan wilderness.

COP NIGHT FIGHT, AUGUST 2009

Whang… Boom! The sound of a crashing 107 mm rocket is something you never forget. That scream is forever etched into your brain.

I heard the noise while lying in bed. Glancing at the clock, I saw that it was just after midnight. The enemy rarely attacked at night, and I thought that this was one of the doomed big attacks. I quickly donned my armor and grabbed my rifle. In those ten seconds of dressing, I heard the echoes of battle increase outside the walls of my brick-and-mortar barracks.

I ran outside my barracks and watched as the night sky illuminated with the exchange of tracer fire.

I waded into darkness and yet another fight.

* * *

BATTLES ON THE FRONTIER

Few people will ever hear about the daily fighting that America's young men and women participated in over nearly two decades in Iraq and Afghanistan. Some of the major battles, like Anaconda, Ramadi, and Fallujah, may echo throughout history. However, many battles will go unrecorded and even forgotten, except for the dozens of memoirs written by those who fought there.

In my brigade, some platoons averaged as many as 40 fights a month. My platoon averaged around 20. The severity and duration of each engagement varied. Sometimes, these skirmishes involved a few bursts of machine-gun fire, lasting perhaps 10-20 minutes. Other times, they escalated into pitched battles with RPGs alongside machine gun fire and could last hours and, at times, entire days. We occasionally encountered IEDs, which could rip open our armored vehicles like soda cans. At other times, we faced rockets combined with mortars.

We are familiar with the significant battles that occur during guerrilla warfare throughout military history. They echo to us in words like Khe Sahn, Black Hawk Down, Fallujah, Anaconda, Wanat, and, in my area, COP Keating. Yet, the thing about these counterinsurgency wars was that, unlike conventional war, the threat of death did not come in the culmination of campaigns, beach landings, or airborne assaults.

Instead, in counterinsurgency, death waited for us nearly every day. It was there on every patrol waiting for you. There was no front line. The front was in all things, in the air we breathed and the sand that blew into our faces and eyes. We tasted the war every day. Even something as arbitrary as taking a piss could get you killed. My first time being shot at with a sniper rifle was while walking to take a piss on my COP.

I prepared myself to die before every patrol in Afghanistan. This may sound, at first glance, overly dramatic. But I want to assure you that this is the truth. Each day, I waded into the River Styx.

WALKING AND THINKING ABOUT THE PATROLS

As we approached the trail's starting point, I thought about the Roman soldiers who occupied this wall centuries ago. I wondered how much they resembled us, modern soldiers—how they lived, trained, and fought on this distant frontier. I imagined their daily routines: drilling under the cold northern sky, foraging for food, and preparing for long, lonely patrols through hostile terrain. Just like us, they would have ventured beyond the safety of their fortifications to project force, maintain control, and keep raiders off balance.

I envisioned them preparing for patrols, fastening their belts, and securing their swords and daggers. I imagined their hushed complaints as they hefted their heavy shields and pilums, their helmets shining under the same sun that now beamed down on me.

Then, I thought of my patrol preparations—the weight of my body armor, the habitual chamber check of my rifle, the murmured curses of my men as we adjusted our gear. The centuries between us seemed to vanish. Different weapons, different empires, but the same reality.

DRESSING FOR A COMBAT PATROL

Time never stops. It moves forward, indifferent to fear or fate, pulling us closer to whatever lies beyond the wire. I never knew what a day would bring—none of us ever do.

I sat up in my rack, staring at the red glow of my clock. It was

5:29 am. For a fleeting second, I wished that time would freeze, that this morning would stretch into eternity, protecting me from what was waiting out there. I felt like a kid again back in Michigan, hoping for a snow day to cancel school. But war doesn't have snow days. War doesn't grant reprieves.

I took a deep breath, trying to steady my unease. Phobos, the Greek god of fear, lingers in the corners of my mind. I try to exhale him away. My hands rested folded between my knees, and my eyes locked on the dusty carpet of my room—stained with sand, dirt, and the filth that seeps into everything here.

Fuck. Time to get up.

I sat up and let my feet rest on the cool concrete floor for a few moments. Will this be the last time I feel it? I pushed the thought aside and rubbed the sleep from my eyes. I knew where I was. I knew what I had to do. We had to go out there. We had to leave the walls of our outpost and hunt them before they hunted us.

I grabbed my uniform, shaking loose the Afghan dust from yesterday's patrol. Gray. I still couldn't believe the Army had chosen this color for use in a war in an arid environment. I laced up my hiking boots—the ones that set us apart from the soldiers at the bigger FOBs. Their tan boots are for walking on gravel; ours are for surviving these mountains.

Then came the weight.

I lifted my heavy armor over my head, feeling the strain as I settled it onto my shoulders. The scent of sweat, dirt, and gunpowder clung to it. It will never be clean again. I secure the straps, running through my mental checklist: nine magazines, smoke grenades, NVGs, extra batteries, radio, extra radio battery, first aid kit, tomahawk, map, and map pens. My camelback was filled with two liters of water, along with a VS-17 panel (a bright orange flag used for marking your location). Lastly, my assault pack—more water, extra mags, fifty rounds of 7.62.

This all weighed around 50 lbs. or so.

I strapped on my knee pads, dropped to one knee, and whispered a quiet prayer.

I was ready for the abyss.

* * *

ON THE WALK

We arrived at Bowness-on-Solway. The small streets, buildings, and of course, pubs were a welcome sight. To this day, the charm of European villages never fails to bring a smile to my face. Walking through the narrow streets reminded me of black-and-white photographs from World War II, showcasing urban combat settings.

Our trip in 2010 was just before international cellular data became easily accessible. However, the trail, for the most part, is well-marked. We navigated traditionally, using a compass, my map, and, of course, asking the friendly people along the trail for directions. The locals here seemed well accustomed to foot traffic and were accustomed to temporarily lost hikers. With a smile, they directed us through a small alleyway leading to the Tyne Estuary, marking the true beginning of our march along Hadrian's Wall.

We soon found the trailhead, and it was a welcoming sight. I smiled at the small archway, reading the Latin inscription above: SEGEDVNO MP LXXXIIII FORTVNA VOBIS ADIST — "Segedunum, 84 miles. May fortune go with you." Segendunum was the last Roman fort on the wall in the village of Wallsend. We paused and smiled for a group photo.

Isaac, Seth, and me about to start an adventure of a lifetime.

I was overjoyed and felt at peace. My blood and body tingled with warmth as I communed with the soldiers who had toiled and built this wall nearly two thousand years earlier. I wondered if any of my ancestors were stationed here and helped construct this place. I couldn't help but reflect on my mother's Maltese maiden name, Galea. In Latin, it means "helmet," a fitting symbol of the Roman legions that once marched and fought across this very land. The thought stirred something profound within me as the path beneath my feet and the wind sweeping across the wall carried echoes of a distant past.

I felt an unshakeable connection—not just to the history of the wall but to the legacy of those who may have called it their post, their home, and their duty. At that moment, the wall became more than just a collection of stones in my mind; it became a bridge to my long-living Roman soul, deeply embedded within me.

We headed east, enjoying the easy, flat path that hugged the Solway Firth estuary, its waters shimmering in the warm afternoon sunlight. The estuary is today one of the waterways marking the Scottish and English border, and it also served as a natural boundary for the Roman frontier. The flowing water, framed by vast mudflats that stretched to the shore, made this place a borderland—a liminal space where civilization met the wilderness. The path, the pace, and the conversation were light and easy, as they always are at the beginning of a long hike. The flat terrain felt like a mirage, masking the challenges that lay ahead.

The ease of those first steps reminded me of coming home from war. At first, it felt very easy—just a few flights, some paperwork, and I was thrust back into peace. It seemed simple initially. I thought I could step back onto life's path and keep walking.

But as I would soon find that, on this hike, just as had been the case with returning home, the real challenges lay farther along, hidden beyond the horizon. The trail's valleys, storms, mountains, and the warrior's soul are not visible at the outset. They remain distant, yet inevitable.

I felt at ease joking and speaking to the brothers as we seemed to glide across the landscape. We joked and talked, and it felt like I was fully present with them for a time. While I was happy to have Isaac and Seth walk alongside me, I was not entirely with them. My feet and my boots were still wet from wading in the waters of the River Styx for too long. While they might not have been able to catch the scent, the stench of war still lingered upon me, piercing my nostrils and periodically making my ears ring with the gunfire and explosions thousands of miles east in those dark mountain valleys of the Hindu Kush.

While I have left the fires of war, her scent still clings to me. It lingers like the smell of smoke from an all-night campfire. I now walk between the living and the dead. All those who have hunted men carry this curse. We bear this invisible shadow like the rucksack of combat, full of all the horrors of killing and nearly being killed.

You can never take this load off your shoulders. You can only learn to adjust its straps better. For some men, bearing the load in times of peace becomes overwhelming. For others, they choose to return to the very death they tried so hard to avoid.

JOHN WADE'S FUNERAL, JUNE 2023

I stepped into the small church, the blast of air conditioning providing a brief relief from the oppressive Texas heat. The space was quiet, laden with sorrow. I made my way to the pews and sat next to my brothers—Flo Groberg, Korey Staley, William Stacey, Josh Campbell, Tyrell Richardson, Robert Cooper, and Andrew Troxell.

We stood one by one and spoke about our fallen brother. Wade had been a force of nature—a beast of a man, built like a Viking. But instead of an axe, he had wielded 21st-century weapons.

Then it was my turn.

I didn't expect it to be this difficult. But as I stood there, my throat tightened. My voice faltered. Tears filled my eyes and ran down my cheeks as I struggled to steady myself.

I shared a story... We had been conducting a hazardous dismounted patrol, moving along the mountainside and clearing the rocky terrain. I wanted to push us farther—extend the patrol beyond the reach of our vehicles and the cover of our heavy weapons. Sgt. Wade listened attentively to my plan. Then, in his steady, no-nonsense manner, he convinced me to reconsider—to keep our march within the range of firepower from our MRAPs.

"Hey, LT... just so you know—there is no reset button."

His advice saved lives. We needed that support later.

I barely managed to finish the story. My eyes scanned the room—the faces of my comrades, the grieving families.

I swallowed hard.

I hoped—prayed—that this would be the last funeral I attended because of a suicide.

* * *

We continued our walk. I loved being in the brothers' company, but a part of me wondered if I was ready to be around anyone. Years later, Isaac, who went on to study psychology and become a therapist, summed it perfectly when he said, "Seth and I were not on the same walk as you." In that single sentence, he captured the unspoken truth of our journey.

Warriors should be cautious about rushing themselves back into the land of the living. Some of us may never fully rejoin them. Sometimes, our warrior spirits drift like ships without sails upon the seas of peace.

I know I am different from other men. I try desperately to fit into society. I coach kids' soccer and lead them in Cub Scouts. Yet, I struggle when I talk to the other men who have not been tested in the fires of combat. Their worries, lives, and jokes are different from mine.

We continued our journey eastward, following distinctive, white-acorn trail markers as we followed a paved road along the marshland. At first, the path was lined with hedges, but soon the landscape opened up to an endless expanse of farmland stretching to the south. To the north lay the channel leading to the River Esk and the River Eden, whose winding shores guided us toward Carlisle.

Open ground. No echoes of war. Here is peace. Here is calm. A pitter-patter of hiking boots and the chatter of friends. No echoes of war. Not a mountain in sight. I am free from death. I am free from war.

The waterway of the Solway Firth along the trail.

There were still no signs of the wall, just the winding path through green fields in front of us. We pressed on, adjusting our packs and stopping every 60 to 75 minutes to rest and care for our feet. I used the method I learned during our many forced marches in the Marines.

During breaks, I spread out my large, green Army poncho for us to sit on. It became our ritual: lying together, changing socks, and liberally applying foot powder. This small act of self-care further connected me to the experience of the legions who marched this same route nearly two millennia ago, their heavy packs probably cutting into their shoulders just as ours did now. Seth, by far the largest man in our group, stated, "I think I'll always remember these breaks."

The path led us through the charming village of Port Carlisle, its stone cottages standing as quiet sentinels against the passage of time. The air was thick with the scent of damp earth and the faint saltiness of the nearby marshes. As we passed the village, the open countryside unfolded before us — fields dotted with grazing sheep and cattle, bordered by hedgerows that swayed in the gentle afternoon breeze. I was especially delighted to see newborn lambs frolicking in the green fields, some of which came close to us. The new life brought a smile to my face.

The march was beginning to wear on us, the weight of jetlag settling into our bones. We found a peaceful spot on a gentle rise just south of the road, west of Boustead Hill, and took a more extended break. The high ground overlooked a patchwork of farmland, its greens and yellows glowing softly under the fading sun. I glanced toward the horizon over the estuary and tried to imagine Roman vessels patrolling this ancient waterway. Stretching out on the poncho, we closed our eyes for what felt like only a moment. We must have napped for just half an hour or so. But a half-hour break can feel like a full night's rest when you're on the move.

Burdened by jet lag mixing with fatigue from the day's hike, we continued our journey toward Carlisle. We soon entered the village of

Burgh by Sands, once the location of a Roman fort. The term "burgh" is the Old English word for a fortified place. I paused briefly to rest by the statue of Edward I, who died there in 1307 while campaigning against Edward the Bruce. I smiled as I realized that these border-lands continued to serve as battlegrounds for centuries, long after the winds of history swept away my cherished Romans. The statue stood there in the fading light, embodying the enduring spirit of these lands, which were always contested and always on the edge of empires. Some places on Earth are those arenas that people continually claim and will fight for with their lives.

We were all starting to succumb to our jet lag. Tired, hungry, and ready for a break, we searched for a pub or bed-and-breakfast around 7:30 p.m. This was an enjoyable time in travel, before the convenience of smartphones with international data plans. Instead, we relied on more traditional means, like asking locals for help. In the village of Beaumont, a friendly man pointed us toward a pub called the "Drovers Rest," located in the nearby village of Monkhill.

DROVERS REST

The Drovers Rest was everything we needed after a long day. The bar felt warm and inviting, with its wooden walls and cozy atmosphere offering a welcome sight for our tired feet and weary bodies. We were eager to rest after marching nearly thirteen miles immediately after our transatlantic flight. The bartender greeted us warmly and, noticing our exhaustion, offered to prepare food for us, even though it was well past his usual cooking hours.

Before long, we dug into plates of hot fish and chips, their salt and grease precisely what we craved. We washed it down with pints of lo-cal beer — Bitter, Cambrian Ale, and Lakeland Ale — each sip a taste of the region and a small comfort after a long day.

As the evening progressed, we debated whether to set up camp for the night or find proper accommodations. A local woman overheard our conversation and offered a solution. She had a friend who owned a nearby bed-and-breakfast; before we knew it, she had arranged lodging for us. The kindness didn't stop there — the bartender, seeing how tired we were, offered to drive us the two miles to the B&B, which was called the Vallum House.

We settled in for the night, thankful for the break and the unexpected generosity of strangers. The weariness of the day melted away, replaced by the simple joy of good food, good beer, and the kindness of those who made us feel at home so far from our own.

King Edward I.

The Drovers Rest pub.

CHAPTER SEVEN

DAY 3, WEDNESDAY, JUNE 23, 2010: CARLISLE TO CAMBECK

Eyes open, I am at peace. A soft and reassuring blanket of quiet surrounds me. There is no fear or lingering stress. For the first time in ages, I dare to think that I might be free. Maybe the dragons in my heart have finally fallen silent. Is that it? Maybe I am ok… Maybe I'm already normal.

I lay awake in the comfortable bed at my B&B, hearing nothing but the chirping of birds outside on the cool summer morning. I rolled over and welcomed the warm sun pouring through the curtains. I was alone in my room, with the two brothers next door. I slowly got up, moving gently and savoring the stillness of the morning. As I sat on the edge of my bed, my eyes settled on my grey Scarpa Kailash GTX hiking boots by the door of the small room. I remembered the day they were issued to me in Afghanistan. They were a welcome addition to my gear, as they were built for the unforgiving mountains that tore through our standard Army boots as easily as the land itself tore through us. These boots served as a reminder: some places are not meant to be conquered, no matter how many empires try.

Afghanistan didn't just shred our boots. It shredded our vehicles, our bodies, our minds.

Boots… I carry him. I carry his load, weapons, water, food, and Armor. I am always with him. I have felt his fear, anger, and sadness. I was there in every fight, every IED, and every incoming round. I was always there. I carry the stench of war forever upon me.

The regular—the normal—didn't survive at war. Like those tanned Army boots, the ordinary was left behind, unable to withstand the brutality of that land. To survive in those rugged places, we had to become something else. Stronger. Harder. More savage. My boots told the story: even the Army had to adapt to walk on that land. We changed in the attempt to conquer the unconquerable. Armies evolve, clinging desperately to survival in places that defy it. Some lands cannot be occupied. They leave scars on boots, on bodies, on souls.

Perhaps this is why my Romans chose to do the unthinkable: wall off part of Britain. While their legions were devastatingly strong in battle, mountainous terrain can bring even the most colossal war machines to a grinding halt. Perhaps this wall was their solution—a means to contain the highlands and their defiant people and to protect the larger Roman cities to the south more effectively than a dozen small, daily patrols from distant legionary camps ever could.

In the West, we often strive to exert and expand our control, to defeat adversaries in decisive, pitched battles where our strength can be showcased. But what occurs when the enemy declines to offer us that opportunity?

During our time in Afghanistan, we attempted to fight differently. We sought to win hearts and minds, investing countless resources, lives, and effort into this strategy. Yet, just when we thought we had made progress, our adversary slipped back into the shadows of the mountains. Perhaps some places will forever prefer their own way of life over ideals, systems, and governments introduced by invaders. Maybe the Wall was the better idea after all.

THE WALK

I finished dressing and triple-knotted my hiking boots, just like I had during countless combat patrols in Afghanistan. The familiar ritual felt like muscle memory, grounding me as I gathered my gear. Once I had my kit together, I made my way to Isaac and Seth's room.

"Hey, I'm going to get breakfast. You guys almost ready?" I asked.

"Yeah! We'll be down in five minutes."

Tea in solitude—few things compare to the simple pleasure of a freshly brewed cup in the early morning before a hike. The warmth seeped into my hands and joints, waking me as the gentle jolt of caffeine reminded me that I still had miles ahead. It was a quiet, restorative moment, a brief pause before the day's march.

The host of the bed and breakfast approached our table, offering a full English breakfast.

"I have no idea what that is," I admitted.

"Try it, Antonio—you'll like it," Seth said with a knowing grin.

He was right. I found myself particularly fond of black pudding—a type of blood sausage. It was rich, savory, and oddly comforting. Sitting there with Isaac and Seth, enjoying breakfast together, was one of the most normal things I'd done in a long while. It felt almost like old times, like those mornings back in Ypsilanti when we'd grab breakfast after a long night of drinking. It struck me how even the simplest acts can feel strange after being away from them for so long.

Before long, we were back on the march, heading directly east toward Carlisle. The plan was to rejoin the path just beyond Carlisle Castle. On the way, I made a quick stop at a small one-stop shop to pick up some supplies. In hindsight, I should have bought more—it's amusing how little you think you need until you're miles from the next opportunity.

Carlisle was a beautiful city, offering my first glimpse of urban life in the UK. The medieval castle rose prominently, its formidable stone walls brilliantly blending with the modern cityscape. It served as a striking reminder of Carlisle's long history as a frontier city, where the need for security has been constant through the centuries.

Long before the Normans built Carlisle Castle in 1092, the site had been home to Luguvalium, a Roman fort established in the 1st century CE. As I stared at the castle, I thought that Luguvalium must have played a crucial role in guarding the northern frontier of Roman Britain. The old Roman fort was long gone, and the land had been reused for the medieval castle.

As I gazed at the deep red rocks of Carlisle Castle, I discovered that its striking color arose from the red sandstone quarried nearby. Yet, while standing there, I couldn't help but think the red symbolized more than just the material—it seemed to embody the tremendous amount of blood spilled in this ancient borderland. This was a place where the cultivated, civilized fields of Roman Britain clashed violently with the wild, ungoverned highlands of the north.

The brothers and I walked in the shadow of history, tracing the blending of Roman and Norman power. The towering Carlisle Castle above us, symbolized centuries of conflict and fortification, serving as a testament to the perpetual attempts to impose order on the North.

As we moved, I felt the shadows of yet another indomitable frontier pressing down on me—those of my own highlands in the Pech River Valley. Unlike the enduring Roman walls and Norman castles, the fortifications we built were ephemeral, quickly consumed by the rugged land we tried to control. We didn't construct fortresses to withstand the ages; instead, we built small, temporary outposts of barbed wire, HESCO barriers, and sandbags—fragile defenses against a land that could not be tamed. And, unlike the Romans and Normans, we left little behind when we departed. No grand walls, no lasting marks—just memories of the dead, those we killed and those we lost, in the dust of an unyielding frontier.

We paused briefly to take photos, trying to capture the castle's imposing presence amid the city's vibrant energy. Nearby, we stopped at the Sands Centre to use the facilities and enjoy a quick rest. Some locations along the trail offered commemorative stamps for those carrying a passport-like book or journal. I couldn't resist. I proudly added a Hadrian's Wall stamp to my journal, a small but meaningful keep-sake of my journey so far.

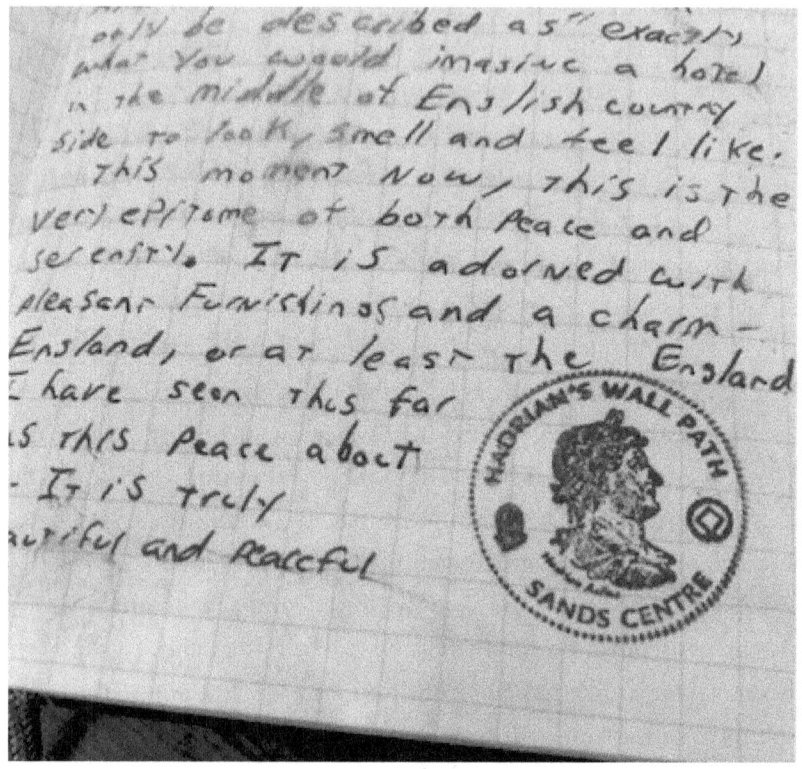

My wall journal

Seth and me in front of Carlisle Castle

While it was nice to walk and explore the city for a while, I found myself yearning for the countryside. I had had my fill of walking on concrete—and perhaps my fill of civilization as well. Behind Carlisle Castle, we rediscovered the familiar white acorn trail marker and rejoined the Hadrian's Wall Path. Not long after, I was captivated by the sight of the slow-moving waters of the River Eden.

This river, as recorded by the Roman historian Ptolemy, bears a name that probably traces back to the Celtic word "Ituna," meaning "water." Its presence seemed timeless as if it had always been there, connecting the past to the present.[7]

As we walked and chatted, I fixed my gaze on the river. Rivers

---

7. Robert Ferguson, The River-Names of Europe (London: Williams & Norgate, 1862) https://www.gutenberg.org/ebooks/35900.epub.images.

endure far longer than any man-made border, I thought. For the Romans, the Normans, and the English, rivers like the Eden helped to keep the northern tribal peoples at bay, serving as natural boundaries against chaos. They were shields, keeping the darkness out.

As I gazed at the calm waters here, I couldn't help but think of my river—the Pech. I remember the first time I saw it during my brief stay at COP Michigan in the early days of my deployment. From the watchtower, standing next to a soldier operating a 240B machine gun, I watched the cool river wind its way around our fortress, a rare moment of tranquility amidst the mountains. Throughout my time in the Pech River Valley, especially during foot patrols in the relentless 105-degree heat, I often looked at that river and yearned to escape into it. I envisioned stripping off my gear and diving into its waters, just as I had in more uncomplicated, more carefree days as a boy. Yet, that river was no sanctuary for peace in the shadow of the mountains and the specter of war. It was a boundary, a lifeline, and at times, a curse.

We walked across the Memorial Pedestrian Bridge, crossing the River Eden and leaving the city of Carlisle behind. In its place, the extraordinary charm of the English countryside unfolded before us, basking in the gentle warmth of an early summer day. A cloudless blue sky stretched above, a blessing for our journey, as we passed through lush wheatfields—the Roman Empire's lifeblood and still one of the world's most vital crops. I smiled at the sight, struck by how the rhythm of the land remained unbroken, echoing through millennia.

Our path carried us through rural landscapes and quaint villages, each more picturesque than the last. Villages like Rickerby, Linstock, and Crosby-on-Eden welcomed us with their timeless beauty, a blend of pastoral serenity and quiet history. For lunch, we stopped at the Crosby Lounge, a building constructed in 1802 in the heart of High Crosby. Although the chefs were off duty, the housekeeper graciously offered us sandwiches, drinks, and seating in a beautiful dining room.

The River Eden

The atmosphere was everything you'd expect from a countryside inn — warm, inviting, and rich in the charm of a bygone era. As Isaac and Seth sipped tea, I chose a beer. We settled into the lush accommodations, resting our feet and bodies in a moment of peace. It wasn't just the physical comfort of the plush furnishings or the generous hospitality; it was the undeniable serenity of this place, a stillness that seemed woven into the very fabric of the English countryside.

Sitting there, surrounded by soft light and simple beauty, I felt a rare calm wash over me. England—or at least the England I'd seen so far—had this quiet power to slow time, soothe, and remind me of life's gentler rhythms. It was beautiful. It was peaceful. And for this fleeting moment, I felt perfectly at ease.

A well-deserved lunch

The trail became even more enjoyable, weaving through the lush green countryside with a rustic charm that seemed untouched by time. Our journey continued further east, passing through the town of Newtown in East Cumbria. With each mile that passed, the number of fellow travelers dwindled until hardly any were left. I cherished the solitude. The quiet of this place felt restorative, its peace a welcome contrast to the noise that so often fills the mind.

The further east we went, the more serene it became, but the solitude also gave me time to think—perhaps too much time. Memories and thoughts stirred in the stillness, unbidden yet unavoidable.

Sunset comes late in the English summer, casting a golden glow over the land long after what I was used to. Even as the day faded, we had enough light to walk well past 9:30 p.m. The words of Tacitus came to mind, beautifully capturing this strange phenomenon: "If no clouds block the view, the sun's glow, they say, can be seen all night long: it does not set and rise but simply passes along the horizon."[8]

Here, it was never truly night; the faint glow of the sun lingered just above the horizon. The effect was otherworldly, lending an eerie beauty to the Roman frontier—and, in a way, to the frontier of my own soul. The endless twilight seemed to blur the line between day and night, between peace and unease, echoing the strange space I felt within myself.

The toll of the march was beginning to weigh on Isaac and Seth more than it did on me. Not that they were out of shape—they weren't—but their endurance wasn't at the same level as mine. For the past year, my daily combat load in Afghanistan consisted of a helmet, body armor, ammo, water, radio, weapon, and map kit—a combined weight of about 50 pounds. For more extended patrols, I carried an additional 10 pounds of water, food, and batteries. That constant strain had forged me into one of the best physical conditions of my life.

By the time we began this hike, I was already accustomed to long days under heavy loads. In my final two months in Afghanistan, serving as an intelligence officer, I often ran 8-12 miles in the mornings and capped it off with a few intense jiu-jitsu bouts before starting my shift. The routine kept my body conditioned and my mind sharp. In comparison, the weight of my hiking pack now felt light, even refreshing, and the march was more meditative than grueling.

---

8. Tacitus, *Agricola and Germania*, trans. Harold Mattingly (Harmondsworth, UK: Penguin Books, 1948), 10.

That evening, as the light began to fade, I felt the urge to press on a few more miles. There was something about the rhythm of the walk, the fading warmth of the summer sun, and the quiet determination in my steps that pushed me forward

"Hey guys, it's only 6:30 pm. We could get another six miles or so in before sundown if we push," I said, trying to sound encouraging.

"Come on, Antonio... we'll make it up tomorrow," Seth replied, his voice a mix of weariness and practicality.

I paused, calculating the distance in my head. A gnawing worry started to creep in — I wasn't sure we would finish the hike across the island in the time we had set. Completing this hike wasn't just a goal for me — it felt like a necessity. Yet, I recognized we were all on different journeys, each of us carrying our burdens and moving at our own pace. My anxiety about the timeline was mine alone, and I chose not to share it. We were far from any village or hotel, and the brothers looked exhausted. I smiled and agreed to start looking for a place to set up camp.

We discovered a small forest nestled along the trail just northeast of Newtown. The trees formed a quiet, protective canopy while a creek named Cam Beck flowed nearby. Its gentle babbling provided the perfect soundtrack for a peaceful evening. It was one of those campsites that felt like it had been waiting for us — perfect in every way.

With a collective sigh of relief, we dropped our packs and began to set up camp. Seth and Isaac took charge of the tent while I gathered firewood. Collecting kindling felt grounding, a small but purposeful task that allowed me to release the day's worries, at least for a little while. We found a moment of peace in this serene corner of the trail, surrounded by the quiet blessings of nature. We set up camp next to a gentle stream. Just beyond the trees was a field peppered with cows and sheep, their presence signaled by the occasional lowing and bleating. Their sounds drifted through the stillness, blending with the soft murmur of the stream and the crackle of our fire.

The scene felt perfect, as if it had been pulled straight from a dream. I sat by the fire, journaling about my journey—this dream I was living. It struck me that we had pitched our camp just meters from where an ancient Roman fortification once stood. Amazingly, here I was, at the very edge of what was once the Roman frontier, sitting where Roman soldiers likely gathered millennia ago.

The three of us engaged in small talk, our voices light and unhurried in the tranquility of the evening. Isaac and Seth appeared visibly exhausted from the day's march, but I still felt strong, almost invigorated. As they rested, I fixed my gaze on the fire, its dancing flames reflecting the flickers of thought in my mind. The flames transported me back to my war, reminding me of one of our trucks engulfed in flames.

JULY 2009

In the bright orange flames of the campfire, I saw many things: my heart, my warrior soul, my pain. The smoke rising reminds me of my war. The fire reminded me of one of our trucks engulfed in flames.

WHANG! WHANG! The impact of a 107mm rocket crashing into the ground near me injected a true serum of fear into my veins.

I was enjoying a rare afternoon nap following our early morning patrol. Leaping out of my rack, I stood up for a tenth of a second. The incoming alarm blared that god-forsaken siren.

Quickly, I donned my heavy armor over my head, snapped on my helmet, and grabbed my rifle. I left my room and looked out the hallway door, which led outside. I ran to the door. As I ran, I could only think about how much I did not want to go out there.

Another round landed as I ran. This could be it.

In attacks like this, we never knew if what we were experiencing was just harassment fire or the prelude to something worse. I paused momentarily at the door, my hand hesitating on the handle.

I took a deep breath and ran as our mortars replied in the distance. The TOC was only 25 meters or so away, but it felt like a mile. I sprinted.

I made it to the TOC. The commander and a few soldiers scrambled to determine where the enemy might be. I knew they were in the mountains, but the mountains surrounded us.

There were more impacts—some WHUMPS…. mortars. The world was on fire. I ordered my men to get to their vehicles and mount up. We might need to hold the ground.

Then another explosion blew the wooden door to the TOC nearly off its hinges. This round blew up our communication antenna. Damn, this might be it. I looked at the men's faces around me. The fear was spreading quickly.

In a tenth of a second, I remembered reading All Quiet on the Western Front and its description of being under heavy bombardment. I was deciding where the best place to be was. Should I stay in the TOC? Should I get to my vehicle

Then I heard the WHANG! But a different sound with it.

"Sir! They just blew up our 47 truck!"

That truck was next to Tower 4. The soldier in the tower screamed. They were targeting us! Fuck!

The 240-machine gun rang out. "I know where they are at!" the soldier screamed over the radio. "We need someone to get here. And call in a fire mission."

That request echoed through the TOC. I made eye contact with the artillery officer, who was covered with dirt, having nearly been killed by a recent explosion. He looked at the ground after meeting my gaze. He needed at least a few minutes to recover.

I took a deep breath and realized that I needed to get out there and call the fire mission myself.

I ran outside to make the 150-meter sprint to Tower 4. The air was on fire, and black smoke from our damaged buildings and vehicles rose toward the sky.

"WHAANG!" Another rocket impacted.

"Are you fucking kidding me!" I said to myself as I threw my body to the ground.

I still had another hundred meters to go. About twenty meters away, dust kicked up from an incoming mortar round. I got up quickly and resumed my sprint.

"Don't die… don't die… don't die… don't die… don't die." I said to myself as I ran.

The soldiers in the tower saw my approach and increased their rate of fire. I made it. I climbed the steps and looked at the smoking hulk of metal that used to be one of our MRAPs. The steel was so mangled that I couldn't help but wonder what my own body would have looked like if a rocket had impacted near me. The fear, now mixed with adrenaline and duty, formed a strange grog in my stomach. It was hard to ignore Death when she was so close to me.

The tower's 240-machine gun roared and violently sent shock waves through my body. Goosebumps and chills ran down my back from the thunder emitted in the confined space of the concrete bunker-like tower. My soldiers looked at me for a moment and said, "Good to see you here, Sir."

What he really meant was, "Thanks for joining us in our moment of need. If we are to die, may we all die together."

"Show me where the enemy is!" I shouted.

They pointed the way with tracer fire. I was ready to bring death on them.

"Mortars, Dagger 46, adjust fire Grid over."

"Adjust fire grid out."

I watched as our own 120mm mortars destroyed our enemy and permitted us at least another day of life.

* * *

I took a moment to gaze into the flames of the campfire, their dance both chaotic and soothing in the night. The subtle crackling of the fire mingled with its scent, and for a fleeting moment, I felt at peace. Nearby, the gentle baa-ing of sheep blended into the soundscape along with the fire's noise. The scene was almost too perfect, and I wished it could have lasted forever.

As I gazed into the flames, I felt myself drift, losing touch with the present and connecting with those long-lost Roman souls sent to their empire's frontiers two millennia ago. Like them, I was a soldier far from home, staring into the fire, searching for answers, and contemplating destiny and fate. At that moment, on the frontiers of Rome, I was beginning to find myself.

The journey drew me deeper into my thoughts, peeling back the layers of armor I had built around my soul. In my mind, I could almost hear the horns of Briton warriors sounding just beyond the wall, calling out across time. This wall and boundary between two worlds — the Roman and the barbarian — mirrored my existence after the war.

I walked between two worlds. One foot was firmly planted in the River Styx, in the land of battle, chaos, stress, and death. I still smelled and tasted my war. I still carried its fear, a tight knot that curled in my belly, refusing to leave. Even now, as I write, that fear still lingers, uninvited but present.

Writing about the war resembles stoking the embers of a campfire in the morning. At first glance, the fire appears extinguished, reduced to cold ash. But with a bit of prodding, the flames revive, hungry and vivid. As I write, those flames rise, carrying with them the memories — the sounds, the smells, the rawness of it all. I gaze back at them now. I remember.

In the fire, I saw my soul. Each lick of the flame seemed to whisper the story of warriors across thousands of years, a chorus of hopes and dreams, struggles and triumphs. War does the same — it's always with us, a silent companion, refusing to be forgotten. It was with me there on that wall, and I didn't even realize it.

But beneath the calm surface, the anger still smoldered. It was deep within me, unseen but growing, beginning to tear at my soul like an ember, waiting for the wind to fan it into a blaze.

"You okay man? You're being kinda quiet," Isaac asked, his voice cutting into the night. Our eyes locked for just a moment.

"Yeah, I'm fine," I lied.

Yeah... I'm fine. I lie because I can't tell you what I mean. Can I tell you, that I already miss war? Can I tell you that I feel like a stranger, even with you guys?

The truth is that I am not ok.

The truth is I might be just a little bit fucked up still.

As much as the screaming of the bullets, the impacts of artillery, and the screams of men haunt me, they tempt me. Sadly, I think that they always will.

I miss being a god, I miss being a warrior.

That night, I felt the darkness stirring within me. I wanted to push the pace, keep moving, and take charge like always. But here, I wasn't a commander. I wasn't in control, which gnawed at me more than it should have.

Maybe that's what happens to us. If we're not careful, if we don't hold the line, we lose control.

CAMBOGLANNA CASTLESTEADS

Many years later, I discovered that we had unknowingly camped near the Roman fort of Camboglanna, now known as Castlesteads. In Brittonic Celtic—the ancient language of the Britons, which evolved into Modern Welsh, Cornish, and Breton—the name means "bent valley," referring to the curve of the nearby River Irthing.[9]

Of course, we had no idea at the time. But perhaps the Roman

---

9. https://en.wikipedia.org/wiki/Camboglanna.

surveyor who first broke ground here noticed what we did—a place that simply felt right. It seemed fitting that on our first night camping along the Wall, we had stumbled upon this spot by sheer chance. This discovery added a touch of magic in hindsight to my memories of the adventure.

CHAPTER 8

DAY 4, THURSDAY, JUNE 24, 2010: CASTLESTEADS TO GREENHEAD

I woke as the sun began to kiss the horizon, its first rays brushing the sky with soft strokes of gold and orange. The crisp summer air greeted me, wrapping my body in its cool embrace. I wasn't sore—my body, conditioned by years of hard marching and physical exertion, hadn't yet succumbed to the demands of the trail. I sat up slowly, savoring the morning's peace, and carefully grabbed my boots, mindful not to disturb my sleeping friends.

There's a pure serenity in waking up in a forest and feeling quiet seep into your soul. Outside the tent, I stood still for a moment, smiling as the sunrise danced through the trees. Warm orange light filtered through the branches. In the distance, the sheep and cows announced the start of their day with incessant mooing and bleating—a peculiar harmony that made me smile once more.

I moved to the fire, stirring the embers and adding a small log to coax the flames back to life. The warmth flowed into my body, preparing me for the day ahead. The scent of smoke filled the air—a perfume as old as humanity itself—and I closed my eyes for a moment, letting its familiarity calm me.

But even in this peaceful moment, my mind raced. I couldn't stop calculating our pace, the miles ahead, the time we had left. Anxiety gnawed at me, a quiet but persistent whisper at the back of my mind. Will we finish in time? I glanced at the tent where Isaac and Seth were still sound asleep. My watch read 6:00 a.m.—they wouldn't be ready for hours. If it were just me, I could already be on the trail, possibly covering 9.5 miles before 10:00 a.m. Every part of my warrior soul wanted to finish this wall, to conquer it mile by mile. But doubt crept in: Can Seth make it? Can we all make it in the time we have?

I sighed and stared back at the fire. An hour later, I heard the soft rustling of the tent as Seth woke. He walked over to the fire, rubbing the sleep from his eyes. I smiled.

"Good morning, Antonio. How are you feeling?" Seth said.

"I'm good. How are your feet?" I replied

"They're sore," he said.

Seth and I followed the soothing sound of a burbling stream. We knelt by the bank and splashed the cool water on our faces. It was such an innocent moment, so pure and straightforward, that it pulled me back to my childhood. A cascade of memories rushed over me, unbidden yet welcome, of the small creek near my childhood home where I'd spent countless hours catching fish, crayfish, and garter snakes. I could almost feel the muddy banks under my feet and the thrill of discovery as I squatted by the water's edge, hands reaching into the unknown. Those were carefree days, untouched by the weight of responsibility or fear.

After a Spartan breakfast of granola bars and bread, we broke camp and continued our eastward march. The pull of the wall grew stronger with every step, tugging harder at my soul as if it were calling me onward. Shortly after passing east of the town of Walton, the landscape began to change. The flat pastures we had grown accustomed to gave way to hills that rose steadily before us. The first of these was Craggle Hill, a 134-meter climb that, while challenging, was a mere prelude to the rough terrain ahead.

As we ascended, our breathing became more labored under the

weight of our packs. The trail was no longer the gentle walk it had been; it was becoming increasingly rugged, demanding, and steep. This change clarified why the Romans chose this location to build their wall. They were masters at combining the natural strength of the land with their incredible feats of engineering, turning hills and cliffs into impenetrable fortifications.

When we reached the top of Craggle Hill, we all paused to catch our breath and take in the view. Rolling hills stretched out before us, framing the ever-closing highlands. It was surreal. In this place, so steeped in history and natural splendor, I felt a faint whisper in the breeze, as if the land itself were saying, "Here, perhaps, a warrior can find peace."

Shortly after our climb, I caught my first glimpse of the wall, and the sight stopped me in my tracks. This place was called Hare Hill, the tallest surviving section of the wall, standing at an imposing three meters high. I walked up to the ruins, my pace slowing as I drew closer. I closed my eyes when my hands finally met the cold, weathered stone. In that moment, time folded in on itself. I could feel the echoes of those who had stood here long before me — warriors and laborers alike. Their struggles, fears, injuries, and miseries seemed to flow through the stone and into me.

Finally seeing and touching the Wall confirmed what I had felt all along: I belonged here. This journey wasn't just a hike for me; it was fast becoming something far greater. Running my fingers along the stone felt like touching the very skeleton of the long-lost Roman Empire. This wall, this enduring monument, had outlived the flesh of the men who built and fought around it, their bodies long returned to the Earth. The stones, worn and weathered, were all that remained of their labor, duty, and lives.

I paused to reflect while standing on what was once the ancient front line. So, this was it. This was where the Roman Empire stopped, where its reach met its limit. I thought about the soldiers stationed

here, thousands of miles from home, gazing out at the wilderness beyond. This wall wasn't just the boundary of an empire, but also the
edge of their known world, the borders of their light. In some ways,
this wall was their Afghanistan—the frontier of their ambitions,
where their conquests came to a standstill. This wall embodied that
limit, a testament to their strength and constraints.

Hare Hill

We stopped briefly at Tower 52a for a much-needed break and to change our socks. Leaning against the cold, weathered stones of the ancient turret and watchtower, I couldn't help but marvel at the history embedded in these walls. Structures like this, constructed roughly every Roman mile (approximately 1.48 km), once enabled legionaries to keep watch over the dark and untamed frontiers. Sitting there, tending to our feet, the weight of history seemed to merge with the physical burden of our packs.

Tower 52 A

I felt strong, my body accustomed to long marches and physical strain, but I noticed the brothers starting to feel the effects of the trail. Their 30-pound packs were taking a toll, and their steps became heavier each mile. Still, we pressed on, determined to continue our journey eastward.

In the early afternoon, as we approached the Roman fort of Birdoswald, the wall began to reveal itself, at least what remained of it.

Small sections, measuring just one to three feet in height, emerged from the lush, green grass—ancient whispers of a once-mighty empire. It ran alongside us now, breathing memories of power and resilience.

I couldn't help but smile as I hiked alongside these ancient ruins. It felt surreal, almost too perfect to be walking here in the shadows of Rome. However, as I continued, my thoughts turned inward and I considered my own country's efforts. How would they be remembered? I reflected on the walls and Hesco barriers of my outpost in Afghanistan—the makeshift fortifications we had constructed in that distant land. What would happen to them when Americans no longer held them?

Even in 2010, years before the final withdrawal in 2021, I sensed that our efforts there would ultimately mirror those of the Romans here. This realization saddened me as we hiked. The wall stood as a reminder of an empire's strength and limitations, illustrating how even the mightiest endeavors are often eroded by time and the relentless currents of history.

Around 2:30 p.m., we arrived at Birdoswald, just west of the village of Gilsland, and stopped for a late lunch. This fort is one of several strategically built, approximately every five miles along the wall, and was part of the intricate network of forts, milecastles, and turrets that formed a robust defensive system. It allowed the Romans to monitor the frontier, respond to threats quickly, and project their power northwards.

Here, we found temporary shelter within the remnants of the ancient fortification, now turned into a museum. We took full advantage of the facilities and had our first hot meal in over 24 hours. The food's warmth and the rest gave a much-needed boost to our strength and morale. As I looked around the site, I imagined the countless soldiers who had once gathered here, seeking rest within its walls just as we did now.

After about an hour, we resumed our march eastward. In the distance, a series of massive hills emerged, looming like sentinels over the landscape. The locals called these formations 'The crags', their dramatic contours shaped by the Whin Sill—a natural dolerite ridge that runs through Northumberland. "Crag" is a fitting term to describe the steep, rugged cliffs and rock faces which the Whin Sill offers. These towering crags rise sharply from the rolling land, creating a natural line of defense that the Romans wisely incorporated into their wall.

But to me, they didn't look like mere hills or cliffs. As I stared at the closely rising and falling crags, they seemed like the clenched knuckles of the Roman Empire—a fist perpetually hammering against the untamed northern frontier. These rocky outcrops pointed defiantly into the northern lands, a visual metaphor for Rome's unyielding strength and resolve. I hoped others would see them as I did—as geological features and as symbols of Roman might. In my mind, these were not merely "the Crags"; they were the "Knuckles of the Roman Empire," shaped by nature and claimed by an empire that dared to push the limits of its reach.

As the afternoon wore on, we pressed eastward and crossed the bridge at Willowford, beneath which the River Irthing flowed in a quiet, steady rhythm. I paused at the bridge, drawn by the river's dark reddish color. I later learned that this tint originated from the river flowing through the peaty soils that are so common in this part of England. Yet, at the time, I imagined another explanation—the blood-stained color of the river was from the fallen warriors who had given their lives in battle on these contested lands. The thought weighed on me, blending the scene's beauty with the haunting shadows of history.

We halted momentarily and I heard an all-too-familiar sound breaking the stillness: the roar of approaching RAF fighter jets. They streaked across the sky, practicing high-speed maneuvers, their sleek forms skimming low to the ground. Civilians might recognize this sound from air shows or on TV, marveling at the power of flight. But for those of us who have hunted men, those echoes carried a deeper meaning. In that instant, the peaceful countryside and the echoes of conflict seemed to collide, stirring memories I thought I had left behind.

The Roman Knuckles

SUMMER 2009

KA KA KA! WHIZ... Snap snap snap snap! The enemy's bullets screamed around us, cracking through the air with deadly precision. The Taliban had their shit together that day. We were pinned down in a streambed, the rocky walls barely shielding us from the onslaught. My men and I hammered the highlands with everything we had—rifles, machine guns—trying to suppress the relentless fire raining down from above.

Fifty meters ahead, another element of soldiers was pinned down, and they had wounded. We had to get to them. But the air ahead of us snapped and hissed with a tornado of bullets. Every step forward felt like daring death to take its shot.

"Fuck it. I'm going to make a run for it." I said to myself.

"Martinez! Mendiola! You're with me!" I said.

We sprinted. Bullets tore into the ground at our feet, kicking up dirt and sending shockwaves through my legs. We barely reached ten meters before diving behind the cover of a terraced field. The enemy was dug into the mountainside, invisible but deadly. If we pushed any further, we'd add more names to the casualty list.

"Bound back!" I shouted.

We moved back to the streambed, with hearts pounding and frustration mounting. I scanned the mountainside, trying to get a fix on their firing positions, but it was nearly impossible. The enemy machine gun team was well hidden, blending perfectly into the jagged rocks. Every time I popped my head up to laze a grid, Death greeted me with a hail of accurate fire.

"Damn," I muttered, clenching my fists. We'd have to try again. I had one smoke grenade left—intended for marking the HLZ—but desperate times called for desperate measures. As I prepared to make the call, salvation came crackling over the radio.

"Chosen 6, this is Reaper 21, checking in."

Two of our jets were on station. "Thank God," I breathed.

Within minutes, they were cleared hot. I heard the unmistakable roar as they screamed overhead, and hell followed.

"Sixty seconds to drop," came the calm voice over comms.

We ducked low in the streambed, bracing for impact. The world seemed to hold its breath before the ground erupted. **BOOM! BOOM!**

The earth shook violently, shockwaves rattling through every bone in my body. Dust and debris filled the air as the mountain seemed to roar in pain. When the echoes of the blasts faded, the relentless gunfire ceased. At least for now, the enemy had gone silent.

* * *

AN ARGUMENT

The day's march dragged on as the terrain shifted into ever-steeper, relentless hills that seemed to rise endlessly against the horizon. Every step felt heavier, and every breath was more challenging to draw. As a combat leader, I sensed the toll spreading through our group like a virus. Isaac and Seth, unaccustomed to this kind of physical strain, groaned audibly each time they lifted their packs. Even I, hardened by years of combat and endless patrols, felt the weight biting into my shoulders.

It was now 5:00 p.m., and though the sun promised a few more hours of light, I knew we were nearing our limit as a team.

"We can push a little further, maybe until 9:00 p.m.," I said, trying to sound encouraging. "Get in another six miles, set up camp after that."

Seth shook his head, his exhaustion written plainly across his face. "Antonio, we're tired. We'll make it up tomorrow."

It was a fair and logical response. Most people would have agreed. But something inside me rebelled against his words, a spark of frustration flaring into anger before I could stop it. The ember of a rage I thought I had buried suddenly roared to life, spreading through me like wildfire. It wasn't directed at them, not really. It was something deeper, darker—a clawing, suffocating fear that we wouldn't complete this journey. That the end of the wall, which loomed so tantalizingly close in my mind, would remain out of reach.

The pull of that ancient stone was a siren's call, relentless and consuming, urging me to keep moving, no matter the cost. The weight of unfinished business pressed down on me, and I could feel the cracks beginning to form.

Damn... I haven't felt this in a long time. This anger. This fire. I thought I had tamed it, chained it deep inside. But now it was here, and it was growing. My stomach clenched, my throat tightened, sweat pooled in my palms, and my heart began to race.

"*Keep it buried,*" I told myself. "*This is not the time. This is not the place.*"

But then the words slipped out.

"Come on, Seth, don't be *weak*. We can do a few more miles."

Seth's face flushed red, and he snapped back, "Fuck you, Antonio! I'm exhausted. My feet hurt. Don't ask again, or I'll hit you."

His words weren't a threat—I knew that. He was tired and frustrated. But it didn't matter. The damage was already done.

My jaw tightened, and my fists clenched. I stared at the ground, trembling as I tried to contain the tidal wave of rage rising within me.

"*How dare you threaten me? You don't know how I have changed since you last saw me. You don't know what I've done.*"

The thoughts screamed in my head, each one sharper than the last. "*I have hunted our enemy. I have chased Death into the darkness. I have stared her in the face, stared into her empty eyes, and dared her to take me. I have rained fire upon mountains. I have sent bullets, mortars, missiles, and bombs to destroy our nation's enemies.*"

The fire consumed me now, roaring through my veins. My hands shook, my breath came in shallow gasps, and my mind raced. The demons I thought I had left behind in the mountains of Afghanistan were here, clinging to my shadow, waiting for this moment to resurface.

But the truth is, you can't. It is impossible to bury it. You can't run from it. The demons never leave. They follow, haunt, waiting for a moment of weakness to claw their way back to the surface. I closed my eyes and took a deep breath, forcing the anger back into the darkness, at least for now. *But the truth is, it's always there. It never truly goes away. You leave war, but war never leaves you.*

We who have hunted our fellow man are forever marked. Sometimes, we drift through the world like ships without sails, lost in the calm oceans of peace. But deep inside, we yearn for the storm. We miss it. We are the storm.

I stared at Seth, and my pulse thundered in my ears. Hot and unrelenting anger surged through me like a wildfire. My chest tightened, and my hands were slick with sweat. Every breath was shallow and ragged, like I was still on the battlefield, bracing for the next round.

I tried to disengage, glancing down at the cool, green grass. But it was no use. The storm had already been unleashed, and it demanded release.

I glared back at Seth and Isaac. They were exhausted, worn down by the miles and the weight of their packs. And they had no idea of the turmoil they had triggered in me. How could they?

"Damn," I thought. "You don't know what you've triggered, do you? You don't know the kind of storm you've unleashed."

I took a deep breath, trying to steady myself. I knew Seth was not a threat. But the fire inside me didn't care. It burned hotter with every second I stood there. *I can't be around them right now.*" The words spilled out before I could stop them.

"Guys… I can't be around you right now. I'm sorry. I have to go," I said.

The words came out clipped, hollow, but heavy with an urgency I couldn't explain to them. They weren't just words — they were an escape hatch, a desperate bid to keep the volcano within me from erupting.

And then I turned east and hurried away from them at a violent pace. My steps were quick and deliberate, faster with each passing moment. The weight of my pack disappeared in the rush of adrenaline.

I didn't, couldn't look back. I heard their voices calling after me, faint at first, then swallowed by distance. The steepened hills rose and fell like turbulent waves. The Crags, Roman knuckles, loomed ahead, jagged and unrelenting. I pushed harder on the climbs, forcing my legs to keep moving, to outrun the anger clawing at my chest.

Memories crashed over me—hikes through the unforgiving Hindu Kush, the weight of my pack digging into my shoulder and the crack of gunfire echoing in the mountains. They blurred with the present, the hills around me merging with the ones I had once fought to survive. I clenched my teeth and pushed harder, striving to leave it all behind.

After miles of relentless marching, I finally stopped. The silence hit me like a wall. This kind of silence isn't peace; it's an emptiness that echoes with memories I can't escape. My pack thudded to the ground as I dropped onto a rock, my breath ragged and my body trembling. The anger began to subside, but shame crept into its place. My hands still shook as I buried my face in them.

Years later, I'd remember this moment, Seth's voice echoing in my mind.

"Officers don't get PTSD or combat stress, right?" I had once said to him. A lie I had believed.

Here, in these hills, I couldn't lie to myself anymore. The rage, the darkness, the weight of war—it was all still with me. And it always would be. I wasn't running from them. I was running from a storm deep in my own heart.

I had thought I could bury it here on this ancient frontier. But deep down, I had always known the truth: it would be with me forever.

SUNSET AND REGRET

Sunset finally arrived, blanketing a dull orange light over the land. A cool wind swept across the former frontier, I felt guilty and ashamed of my outburst and sorry that I had left my friends behind. Yet, it was precisely because they were my friends that I had to go.

The truth was simple but heavy — I couldn't be around anyone. I needed time to decompress and time to get to know the new me. Like the Native American warrior traditions I had read about, I knew I needed solitude to reckon with the darkness. That darkness still hung within me, silent and heavy as the encroaching night. Alone on the trail, with no hikers in sight, I felt the weight of the highlands stretching endlessly before me. Sadness, hunger, and exhaustion clouded my mind as I marched eastward, accompanied only by the faint rhythm of my gear swaying with each step.

Beside me, the wall loomed — my only companion now.

Walls typically serve two functions: they keep things out. Or they hold some things in. But for me, it served a different purpose. Instead, the wall allowed me to release something. This fire within me burned deep in my soul. This dragon continues to breathe deep within. Here, at the world's edge, I attempted to free something. The wall became my confessor, hearing my confessions as it has listened to warriors for thousands of years. The longer I walked along Hadrian's Wall, the more I realized it was never just about following the trail; instead, it was a journey through my soul.

I walked as we had walked after the war, skirting the thin line between the living and the dead. At times, I wondered if I should have died in the mountains of Afghanistan. Perhaps I cheated Death one too many times, and combat stress, and this lingering guilt — this was my payment to the ferryman. It still is. Death continues to tax us, never satisfied with what we have already paid. She taxed me then, and she still taxes me now. I walked on this spirit road, skirting the land of the living and the land of the dead.

At home, my kids often ask me, "Daddy, why do you have to work out so early in the morning?"

"Because it keeps Daddy strong," I reply with a smile, though the truth is far more complicated.

I rise before the world awakens, not just for the sake of physical fitness, but to wrestle with the demons that claw at the edges of my mind. Their roars are muffled by movement, their chains tightened with every rep, every mile, every drop of sweat. When I stop, when the morning's quiet or the night's stillness creeps in, they return. The visions of shredded flesh, the screams of the wounded, and the deafening crack of bullets breaking the air — they wait in the silence, ready to pounce.

The haunting sounds of battle have lowered their volume in my ears today. Yet they still linger in the shadows of my heart.

As the sky deepened into twilight, I pushed further eastward, and those familiar sensations began to creep back into my body unbidden. My stomach tightened, my throat constricted, and sweat dampened my palms. Doom whispered its promises, threading fear through my veins with each step. I whispered my own desperate prayers into the encroaching darkness.

I continued trudging in the fading light, with growing exhaustion conspiring against me. My thoughts became untethered, drifting in and out of the present. For a moment, I wasn't sure where I was. The energy to read my map or plan my next steps had abandoned me, replaced by the weight of my pack, the echo of the argument, and the shadow of the war. It all bore down on me, threatening to crush my resolve. I pitched my tent at the first patch of acceptable ground I could find. My eyes scanned the darkening forests and fields around me, hoping for some small refuge to appear.

As the last rays of daylight faded, a glimmer of hope emerged — a sign caught my eye: Greenhead Hostel. The promise of shelter and rest called me to the small village. I made my way there, my legs heavy, but my mind heavier. At the check-in counter, the manager greeted me warmly and assured me they had a room available. Relief washed over me, palpable and overwhelming. The room was simple, but it felt like a sanctuary to me.

I hadn't felt this tired since the war. But this wasn't just physical fatigue — the deep weariness of carrying too much for too long. That night, for the first time in what felt like ages, I allowed myself to lay it all down, even if just for a little while.

While having a drink later that night, I started a conversation with a local who was eager to hear stories about Afghanistan. Before going to bed, I decided to wake up early, find some high ground along the trail, and wait for the brothers. When I saw them, I'd apologize and make things right.

Later, I learned that Seth and Isaac had also found shelter for the night at a nearby hostel a few miles west of Greenhead. They, too, had made plans to reconnect with me and to continue our journey together.

CHAPTER 9

DAY 5, FRIDAY, JUNE 25, 2010: GREENHEAD TO HOUSESTEADS

That night, the war revisited me. This time, it was in my dreams.

SEPTEMBER 2009: AMBUSH

My platoon lumbered back in our MRAPs after finishing our patrol. Fatigue plagued us all, and we looked forward to returning to our COP, just beyond the bend of our godforsaken valley road. The air felt painfully quiet, almost too calm, like it always does before a firefight. We drove with our headlights on, as was customary when not moving directly toward contact. While the white beams of light illuminated the dark road ahead, they also became perfect targets.

The gates of hell opened up on us.

BOOM! KA, KA, KA, BOOM!

The night was illuminated by RPG and machine gun fire, raking every vehicle in my platoon.

We returned fire with various weapons, including our .50 caliber, MK19, and 240s. The eerie, laser-like zips of tracer rounds lit up the night sky and the air around us.

As time slowed to a near stop, I looked to the right and noticed a slower, wobblier, laser-like round coming toward us, and I wondered why it moved differently than the others. This round skimmed the top of our MRAP's hood and exploded with a horrific crash just 10 meters from my vehicle. The hellish BOOM erupting from it made it clear that this was an RPG.

Their firepower was so overwhelming that we could not stay near this kill zone. We unleashed our fire, returning shots into the mountain. Three of my four vehicles were damaged, and we could barely escape the kill zone. We limped away from a barrage of fire. The damage to our vehicles would make us a greater liability if we chose to reengage. We called in mortars on the mountain, and we slowly made our way toward our nearby outpost.

Although none of my men died in that close ambush, we had still been defeated. We limped back to the American outpost only three kilometers away in defeat.

"But fuck that," I thought.

I knew if we gave them this victory, their confidence would only grow and perhaps lead soon to our deaths, perhaps if not tomorrow, but soon.

We would return to seek our revenge. Upon returning to the outpost, my commanding officer, Shaun Conlin, ordered us to prepare for our counterattack, along with some reinforcements from nearby COP Able Main.

By that point in my deployment, I had been in dozens of firefights, but a new dread came over me as I prepared for this one, preparing myself to reenter an inferno I had barely survived.

I was in the assembly area of my outpost, waiting to roll out. The company XO, Brandon Fridia came to me.

"Dude, I don't want to go back out there," I said.

The seasoned officer looked me straight in the eye. "Dude, c'mon, you got this." Fridia patted me on the back and gave me the confidence that I so desperately needed.

Within 15 minutes, our reinforcements arrived, and we pushed back to the East to find them. This time, we rolled in the darkness, headlights off. We drove using night-vision goggles, peering into the darkness through the eerie green light of our technology-enabled hunters' eyes.

Returning to the site where we almost died was a perilous feeling. I felt like a firefighter being told to return to a burning house after nearly dying. I prepared my soul to meet God, silently whispering the 'Our Father' as we quietly rolled back to that mountain draw.

As we neared the ambush site, I saw some embers burning on the road from the recent battle. We stopped and scanned the now dark and quiet highlands.

"Sir, I've got movement up there! We don't have any friendlies up there, do we?" my gunner, Cortez, frantically asked me.

"No! We are the only friendlies outside the wire!" I said, wondering why the fuck he would ask me such a stupid question. I turned on the command monitor screen in my truck. It was pitch black outside, but we hunted with night and thermal vision. I saw what my TOW gunner had in his sights and suddenly understood PVT Cortez's question.

It was jarring to see them so plainly—my phantoms, these ghosts, my enemy. The crosshairs on my screen were locked on them, and our TOW anti-tank missile was aiming at them.

The Mujahideen in Kunar were often more men of myth than reality. They breathed fire upon our bodies and our outposts like mythical, invisible dragons. I heard their roars and felt their flames against my flesh. At times, I even saw the fire pouring from their mouths. Yet, they seemed to fly unseen and, for the most part, unheard in the Hindu Kush. They rained death upon us from afar.

Now, it was different. The magical fog which usually concealed my enemy had lifted, and there they were. They walked down the high ground in single file, with their weapons lazily slung over their shoulders, much like how our men carry 240s when not expecting contact.

These fighters had no reason to expect contact. They had just ambushed my platoon, only 20 minutes ago, and won.

They walked with their hearts glowing from a successful ambush. In the brief moment before giving the command to kill, I was mesmerized by the sight of them—dragons who had turned out to be nothing but men. I watched as they navigated the switchbacks down the mountain. I saw their faces glowing brightly through the thermal sights. I could almost smell their sweat and hear them joking as my men would have done after a successful engagement. I imagined the sound the rocks must have made under their shoes as they walked in the darkness.

The guerrillas shone brightly in the dark through our command viewer. The warm blood in their veins made them glow like fluorescent ghosts on my screen.

I felt genuine anger toward them. A difference of a few inches would have sent RPGs melting their way through the armor of my MRAPs and burned my men and me alive. I could have been choking on my blood at the moment, as the Taliban would have mounted our vehicles and carried our weapons away. Perhaps they would have dragged our wounded away and tortured them. Maybe they would have captured the entire experience on video for propaganda.

My blood boiled, and I thought of what I needed to do.

"You're mine now. You fucked with the wrong platoon You thought you won, didn't you? You thought you wounded us and scared us back to our outpost, didn't you? How very wrong you were. We are the Death dealers now. I will show you the truth of battle and the truth of Death. I am Death."

"Take them out now!" I ordered.

I had become Death with just a few words.

BOOM!

A TOW missile launched and spiraled toward my foe. I saw one of the guerrillas take his final breath, gasping as he glanced back at the road. A second later, an explosion turned that squad of proud warriors into enemy casualties.

I no longer feared Death.

I had become it.

* * *

A FAILED LINK-UP

I awoke the following day. The room was quiet, but the guilt of leaving my friends behind pressed heavily on me. It was early—5:45 a.m.—and the first rays of sunlight were beginning to stretch across the sky, painting it in soft hues of pink and orange. I couldn't shake the thought of Seth and Isaac, wondering if they were all right and if I had pushed too hard.

None of us had international cell phone coverage back in 2010. There would be no reassuring texts or calls, no precise GPS coordinates to facilitate a quick reunion. If we were to meet again, it would have to be the old-fashioned way—by sheer luck and timing. I guessed I had put at least a few miles of distance between us. I had to find them.

After a quick breakfast, I laced up my boots and returned to the trail, breathing in the crisp morning air. The world felt alive but empty, populated only with birdsong and the crunch of gravel beneath my feet. I needed a vantage point, somewhere I could watch the path and wait for them.

I knew they were strong and capable men. Still, the thought of leaving them on this ancient frontier tugged at my heart. Within half a mile of the hostel, I came upon the ruins of Thirlwall Castle. The medieval fortification stood solemnly against the horizon, its weathered stones telling the story of time. Built in the mid-1100s, its very foundation bore bricks from Hadrian's Wall, with its stones repurposed from the Roman frontier. Perched atop a gently rising hill, the castle offered a perfect vantage point and a clear line of sight along the trail.

The early light wrapped the ruins in a golden haze, and I climbed to its heights, scanning the winding path for any sign of my friends. It was just past 7:00 a.m. and the trail stretched empty before me. The guilt and the longing to reunite with Isaac and Seth pressed harder now, mingled with a faint hope that I would see their figures cresting the next hill at any moment. Until then, I sat among the ancient stones of Thirlwall, the whispers of the past keeping me company as I waited.

Thirlwall Castle

The hilltop, with its 30-meter rise, loomed before me—a natural fortress. Every soldier understands the power of elevation—it's a gift from nature that transforms defense into an art. The height forces your enemy to labor for every step, their lungs burning and muscles straining just to reach you. Even the slightest incline can blunt the sharpest of assaults. This place reminded me of my own citadel, OP Rocky, in Kunar Province. There, too, we relied on the terrain, the steep cliffs and unforgiving ridges to shield us from the relentless assaults of an unseen foe.

The stones of this castle felt different, though. They held the weight of history. Many had been taken from the ruins of the Roman wall long ago to reinforce this medieval stronghold. I ran my fingers along their surface, sensing the texture of time. Roman hands had shaped these stones, their strength and sweat lingering like whispers. I leaned forward, my breath blending with the chilly morning air, as if to taste the breath of the legionaries who had stood here two millennia ago.

I slipped my heavy pack from my shoulders and let it drop to the ground with a dull thud. I felt relieved, but the strain lingered in my heart. I sat and allowed the morning breeze to wash over me, maybe even helping to cleanse the sands of war from my flesh. My gaze shifted to the trail, scanning every distant movement on the horizon. My heart leapt with each flicker of motion, and I strained to see if it was Seth and Isaac. I desperately wanted it to be them. I longed to see their familiar figures crest the hill, to tell them how sorry I was for leaving and for the anger that consumed me. I yearned to apologize, to explain what I couldn't express before. The regret gnawed at me, a constant ache. I had left my friends in a foreign land, and that truth weighed heavier than any pack.

As I waited, the charm of northern England continued to seep into my soul. It possessed a gentleness and beauty I had never encountered before. I felt as if I were stepping into a medieval kingdom frozen in time. The crispness of the tranquil green landscape refreshed my spirit. This place resembled a storybook come to life. I now understood why this land had been so fiercely defended throughout history. It was something worth protecting, worth cherishing.

But the hours slipped away. The sun climbed higher, and with each moment that passed, my hope of a reunion dimmed. Maybe they had already passed me, I thought. Perhaps they took a bus or veered off the trail. The uncertainty gnawed at my mind. More travelers appeared on the trail—backpackers with sunburned faces, couples chatting in languages I didn't understand, families with excited children—but none were Seth or Isaac. Their absence grew more pronounced with each passing hour.

Five hours passed, taking with them the last flicker of hope. I stared at the ground, my hands resting on the cool stones, my chest heavy with sorrow. They weren't coming. I was sure of it now.

I was alone. Truly alone.

Later, I learned that Isaac and Seth, anticipating my faster pace, had taken a bus to get ahead of me. They waited for hours, scanning the trail just as I had, hoping I would show up. But when it became clear that we had missed each other, they made the most of the situation and continued on to Edinburgh for sightseeing. Their journey took a different turn, filled with new sights and moments of exploration, while I stayed on my own path, grappling with my thoughts and regrets. I was indeed alone.

I held onto a sliver of hope that I might see them again further down the trail. But deep down, I knew they were gone, and the rest of this journey would be mine alone. The loneliness was sharp at the time, but years later, I realized it was exactly what I needed. Sometimes, the meaning of a journey doesn't reveal itself until long after the final steps have been taken.

Combat stress—or PTSD—affects every warrior differently. For some, the impact is immediate, hitting like a hammer blow when chaos ends. For others, like me, it festers quietly, incubating until it can no longer be overlooked. It took me years to realize that it is impossible to hunt your fellow man for months, for years, without being changed by it. The shadow of war leaves its mark on us all.

As I continued my walk along Hadrian's Wall, I began to understand why I was walking. This was not just a physical journey through England's northern frontier—it was a reckoning with myself. That ancient wall became my confessor, my sanctuary, and, in some ways, my battlefield. It was where I first started to confront the war within myself. Those miles represented a cleansing—or at least the beginning of one. It was time to move again. I stood, my body aching from the road, but my spirit a little lighter, and I hoisted my heavy pack over my head. Taking a deep breath, I turned toward the horizon and continued my journey east, further into the frontier of the Roman Empire and deeper into my soul's frontier.

Near Walltown Quarry

THE ROMAN KNUCKLES...

Almost out of nowhere, a series of hills rose sharply from the ground, with the Walltown Crags leading the way. Here, the wall blended with the natural fortifications of the terrain, enhancing its defensive strength. The landscape mirrored my mind's restless, unrelenting, and unforgiving chaos. The rolling hills undulated every 100 to 200 meters, creating a relentless rhythm that challenged my already exhausted body.

Every soldier understands the value of defending a hill—it's a natural advantage. However, any good soldier is also aware of the grueling toll that climbing one can take while carrying a heavy load. With each ascent, my breathing quickened, and my body heat surged.

Breathing quickens. Body heat increases. You wish you hadn't worn that scarf or that sweater. Your quads begin to burn. Even taking a drink of water is an effort.

Here, amidst the Crags, or what I referred to as the Roman Knuckles, I felt both humbled and determined to conquer not only the path ahead of me but also the chaos within.

Hills near Turret 45

ROMAN KNUCKLES

I called these hills the Knuckles of the Roman Empire. Here, Rome struck at the northern frontier, just as my armies had done in Afghanistan. We drove our weapons into these unyielding mountains. We bled, fought, and spent rivers of blood and untold treasure trying to subdue them. Yet, they both remained unconquered. Perhaps some places were never meant to be subdued. The scarred remnants of the wall cling to these hills now, silent testaments to the empire's limits. The mangled knuckles seemed to whisper as I passed: *You have been here before. We have heard your screams. We have heard your battles. We have heard Death.*

This wall had called to me and my restless heart for years, and now I stood among its echoes, continuing the labor of cleansing my soul. Each step up and down these Roman knuckles brought me closer to something—some understanding of myself, some painful reconciliation with who I had been and who I now was. The hills seemed to peel back the layers of my anger and grief, slowly exposing the fragile core of a man attempting to rejoin the world of the living again. It would take longer than I wanted to admit, but I pressed on.

Since I had lost nearly five hours waiting for Seth and Isaac, I knew I needed to quicken my pace to make up for lost time. The knuckles demanded everything I had, their rugged rises and plunges challenging my endurance and determination. Yet, this was my favorite part of Hadrian's Wall. In this harshest region, the wall seemed to blend seamlessly with the crags, a symbiosis of human engineering and natural fortification.

With each climb, my legs burned, and my lungs strained, but my spirit felt a strange sense of peace. Like a loyal companion, the wall silently witnessed my struggle, just as it had for countless others over the centuries. So, I continued east, traversing the undulating terrain, carrying both the weight of my pack and the heavier burden of my thoughts.

Milecastle 39

The terrain challenged me as I crossed the rugged outcrops of the Roman Knuckles throughout the day and into the early evening. Each step over the jagged, undulating landscape reminded me of the enduring resilience required to conquer such a place, whether as a Roman or a solitary traveler like me. I began to slow down as I approached Milecastle 39, one of the most scenic and well-preserved fortifications along the wall. I decided to rest here for a while.

I dropped my pack and rested against the ancient stones, the sun beginning its slow descent into the western horizon. I ate the last of my rations, and I knew I could resupply soon—if not that evening, then the next day. Exhausted and sore from the relentless hiking, I embraced the minor suffering. In these moments, stripped of comfort, I felt closest to my lost Romans.

As I leaned back against the remnants of the fort, I watched the clouds drift lazily across the bright blue sky, their shadows dancing over the rugged hills. The sunlight faded, spilling into the horizon like a golden tide. Sunsets here at the Roman Knuckles were stunning, with long, fiery rays spreading across the ridges like fingers of light. For a moment, time seemed to stand still, and I felt a connection not only to history but also to my warrior soul, rekindled by the quiet struggle of the trail.

Sunset on the former Roman frontier

Eventually, I stood and pushed farther east, the warmth of the setting sun blanketing the landscape in a rich orange glow. The beauty of the moment did little to ease my growing hunger. I hadn't eaten since midday, and the hike had drained me. With a heavy pack, I burned over 400 calories an hour. The hunger gnawed at me, but strangely, it made me feel more alive, raw, and present in the moment. Sweat dripped off my head, and I felt symptoms like a fever begin to creep up on me. I was near my physical limits, and I didn't know how much further I could go. I knew I had to find somewhere to rest soon, but as I checked my map, I noted that the villages were far off the trail.

As the terrain stretched out before me, mostly bare and hilly, I began scanning for a good spot to set up camp. Suddenly, a rise in the distance revealed a small forest atop one of the hills, and my heart lifted at the sight. I felt confident that I could find shelter in that little wood.

The Housesteads Wood in the distance

Mile Castle 37

In the fading light, I came across Milecastle 37, located just a hundred meters or so west of Housesteads Wood. Embedded in the wall was a broken ancient archway, its doorway still standing as a silent sentinel, providing access to the North. This archway had likely witnessed thousands of Roman patrols pass through its shadowed threshold. For the Romans, it was one of the few gateways beyond the edge of their known world. It led into darkness, into the ungoverned and unprovincial lands that defied the Empire's control.

But for me, it was more than a passage into the frontier. Instead, it was a gateway further into the realities of my soul. For those of us who embark on these odysseys from the lands of war, we often straddle two realities: one foot in life and the other in death. Standing there, I paused, peering through the ancient archway into the shadowy forest that stretched toward the rising hills. The Romans had feared places like these. Forests were dangerous to their war machine, whose strength lay in the open ground where the engines of the legions could be fully unleashed. The legions' formations and tactical precision faltered among trees and uneven terrain. These were wild lands—untamed, unpredictable, and hostile to the order Rome sought to impose.

As I gazed into the ancient frontier, I felt a flicker of unease, intertwining with my own combat experiences in the inhospitable terrain of Afghanistan. It was as if the archway symbolized a passage into barbaric lands and the darker corners of human knowledge. It served as a portal to chaos, fear, and survival—the elements of war that still clung to me like shadows.

I lingered for a moment longer, staring into that unknown, but the candle of daylight was nearly extinguished. I knew I had to move quickly to reach the woods and find a suitable place to camp before darkness fully claimed the land. My head now began to pound with a coming fever.

HOUSESTEADS FORT AND WOOD

As sunset approached, I found myself near the small forest and the fort known as Housesteads, or 'Vercovicium' to the Romans. This site was one of the better excavated and preserved forts along the wall, and it was a remarkable experience to walk its perimeter during the eerie sunset. As I circled the fort, I could sense the presence of the Romans. I felt the thousands of heartbeats that came before me. I imagined the murmurs of troops and the faint sounds of blacksmiths at work. I could smell the hay for the horses, and their manure. After finishing my silent patrol of the fort, I returned to the woods.

As I entered the small forest, it was like something from a storybook. The trees provided me with the shelter I was looking for. I was walking on the frontier of life and death. I walked there in the "in-between world." I still do, as those of us who have hunted man might forever do.

SLEEPING ON THE ROMAN FRONTIER

Then I discovered it — the perfect spot. Just beyond the wall, it sat on a small ledge near the edge of a cliff facing north. A few weathered trees stood nearby, like silent sentinels, providing shelter and a guide to frame the view. I climbed down cautiously, feeling the ground shift beneath my boots, and took in the scene. It wasn't large enough for a tent, but it didn't need to be. There was just enough space for my sleeping bag and Gore-Tex sleeping bag shell. That night, I decided to "cowboy camp" under the stars, as I had so many times before — on other mountains, in other lands.

The moment I set foot on that ledge, it felt as if the earth had been waiting for me. This tiny, hidden perch seemed carved by time itself and appeared to have waited through the centuries, just for me. This place wasn't just a campsite; it was a sanctuary. The quiet here felt sacred, and I allowed myself to pause, breathe, and smile. I let the cool evening air cool my head, which was coming on to a fever. I was running low on water, and I cautiously sipped what little I had remaining.

I thought of the mountains of Afghanistan, where I had hunkered down on jagged terrain, waiting for dawn or the enemy to move. Those nights felt different. The stars above seemed indifferent and distant, silent witnesses to the chaos below. But here, under the northern sky, they felt closer and warmer, as if they might listen to my weary soul. I had crossed so many frontiers in my life — deserts, mountains, battlefields — and now I stood on this new one, straddling the line between the warrior I was and the man I was trying to become.

The ledge reminded me of those nights, the places I had called home for an hour or a day in the relentless wilderness of war. In Afghanistan, the mountains were rarely our allies, wearing down our boots, gear, and bodies. Yet here, it felt different. This small ledge seemed inviting, almost as if it had chosen me. I carefully unrolled my sleeping mat, each motion deliberate, savoring the sense of belonging that came with discovering this place. It was almost too perfect, as if the land had decided I deserved this moment of peace after so many years of wandering. Especially since I felt unwell and extremely fatigued, I was overjoyed to have finally found a place to rest.

As I settled in, I gazed at the frontier before me, pondering the Romans who had once stood on these cliffs, looking north into what they called barbarian lands. I envisioned their unease and longing for the Empire's safety in the South. They had their frontiers, just as I had mine. And perhaps, just perhaps, this was my frontier—a place to confront my fears, my memories, and the unshakeable echoes of war.

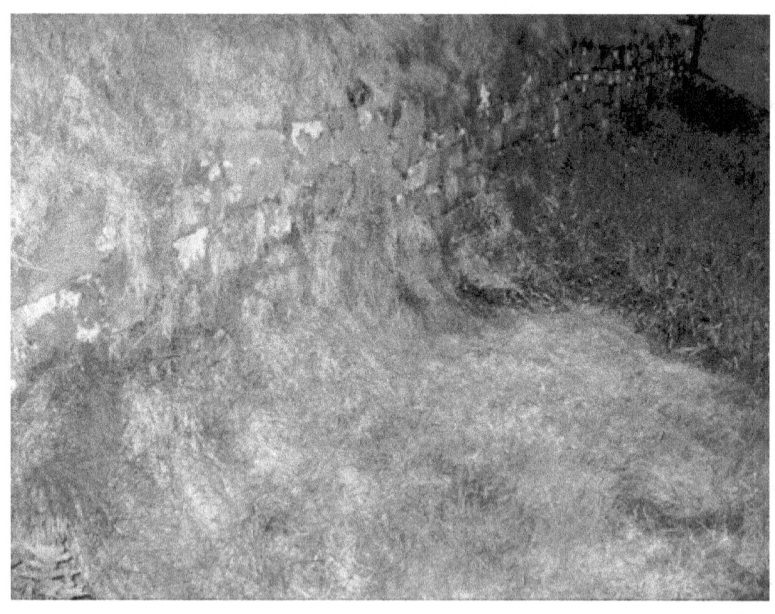

My bed for the night.

I watched the sun slowly descend in the west, its orange light spilling over the highlands like an artist's final brushstrokes. The scene was perfect as the last rays of sunlight streaked across the highlands before retreating beyond the horizon. As the sky shifted to a muted gray, I was alone in the darkness. It felt as if I were sitting on the very edge between life and death—between what was known and what lay beyond.

Little did I know that I had just stumbled upon my favorite place on Earth. It wasn't just the view or the quiet, though both were perfect; it was how this small, unassuming spot seemed to hold me in its embrace, offering a peace I hadn't felt in years. Years later, I still often find myself daydreaming about that moment. When I do, I pull up maps on my phone, zooming in to locate that speck of woodland by the wall. It's as if by revisiting it, even virtually, I can recapture a piece of the solace it provided. I hope I'll always be able to find it easily—a small anchor to a moment of rare tranquility in a life that has seen too much chaos.

Sunset from my campsite in Housestead Wood

I rested, staring into the gradually darkening sky. It was well past 10:00 p.m., yet in Cumbria, the sun stubbornly clung to the horizon, refusing to let go. The sunsets here arrive late; even when the sun finally dips below the earth, it doesn't leave complete darkness behind. Instead, a faint, eerie light lingers, softening the night and making it otherworldly.

The North—these once untamed, wild lands—slowly dissolved into the shadows. Gazing into the dim and unknowable expanse beyond the wall, I felt as though I were staring into the unknown of the Roman world and, simultaneously, into the dark and uncharted recesses of my soul.

I slept there, on the remnants of a Roman Knuckle fortification, where sentries once stood watch. Without modern technology, they would have peered into the darkness, armed only with cloaks against the cold, *pila* gripped tightly in hand, and wary eyes scanning for signs of their foes. I could almost hear their murmurs, shifting footsteps, and the occasional clinking of armor. Beneath me, I felt the heartbeat of an empire that still seemed to pulse faintly in these green hills, even centuries after its collapse.

To hold a fort at night—to fight at night—is to embrace a unique kind of fear that warriors have experienced throughout history. The darkness conceals everything, amplifying every sound, every shadow, and every perceived threat. In the 21st century, with our thermal imaging and night vision technology, we believe we've mastered the night. We think we can see, yet this advantage is frequently exaggerated. Using night vision is akin to illuminating a flashlight in a pitch-black basement—you can only see what the beam touches. Beyond that narrow circle of clarity lies an uninterrupted wall of the unknown.

Lying there, on the same ground where Roman sentries once stood their lonely watch, I felt that timeless fear creep into my chest. The Romans feared what lay beyond the wall—the tribes, the wilderness, the untamable lands. I feared, too—not just the dark of the night but the darkness within me, the things I carried back from the mountains of Afghanistan, the echoes of war that still whispered in the corners of my mind.

This place—this quiet, eerie light, these eternal hills—was a frontier in every sense of the word. A place where civilizations collided, where empires stretched themselves to their limits, and where I, too, confronted my limits.

NIGHT ATTACK,... FALL 2009

I sat alone in my quarters, poring over maps and reports in the stillness of the night. The day had stretched long, yet I was determined to finalize the plans for the next day's patrol. The lamp cast a soft glow over my desk as I tried to concentrate, the quiet scratching of my pen the only sound in the room.

I reached for my journal, the ritual I clung to each night to keep my thoughts organized. But before I could write a single word, it came—an unmistakable sound.

BOOM!!!

The heavy, guttural percussion of an RPG shattered the quiet.

"Damn... that's on the COP!" a soldier's frantic voice echoed down the hallway.

Then the rapid, staccato cracks of machine-gun fire filled the air, followed by more explosions. My pulse quickened as the realization set in. "RPGs on the COP..." I muttered to myself; the words laced with dread. "Maybe tonight's the night."

If they were close enough to fire RPGs, they were close enough to breach. This could be it—the night we'd rehearsed for, feared, and dreaded.

KA-KA-KA-KA!

Our towers came alive, pouring suppressive machine-gun fire to the south.

"Mount up!" I bellowed into the hallway, adrenaline surging. Soldiers scrambled to their vehicles, boots pounding against concrete. My driver appeared, his face tense but focused.

"Get to the truck," I ordered. "I'll be right behind you. I am heading to the TOC first!"

I sprinted into the darkness, fumbling with my night vision goggles as I ran toward the Tactical Operations Center. The night sky flickered with the eerie glow of tracers—red streaks slicing through the black like deadly fireflies. Gunfire at night always felt different. The air seemed charged, electric, alive with the raw energy of danger.

The unknown weighed on me like a heavy pack. Were they charging the walls now? Were they lying in wait? Or was this just harassment, another attempt to exhaust us?

Bursting into the TOC, I barked, "What's the situation?"

The sergeant on radio watch turned to me, his voice steady. "Heavy fire from the southwest, sir. Nowhere else."

"Roger."

I sprinted back into the night, dread gnawing at the edges of my focus. The glowing tracers illuminated the darkness as I sprinted the 60 meters to my vehicle. Another RPG exploded in the distance, the concussion rumbling through the ground like a growl.

"Damn." I muttered, pressing myself briefly against a Hesco barrier for cover. Only 25 meters to go.

"They can't see me," I whispered to myself. The thought steadied my nerves just enough. I resumed my sprint, reaching my MRAP and climbing inside.

"Move toward Tower 4," I ordered my driver.

"Roger that, sir."

Driving toward the sound of guns never gets easier. It's a test of nerves and resolve. Every instinct screamed to turn away from the fight, to retreat to safety. But soldiers don't have that luxury. We suppress that primal fear, channeling it into action.

The world outside as seen through my night vison glowed an eerie green, and the landscape was alien and surreal. Tracers tore the sky as we approached the tower. My gunner, Cortez, scanned the highlands through his thermal scope, his TOW missile launcher locked and ready.

"Hey, sir… you see that?"

"Yeah, I see them. One sec—let me get the grid."

I spotted shadows lurking behind a large boulder through the command viewer. A small team of fighters was testing their luck against us that night. I clearly saw their weapons, glowing hot in the darkness.

I called in a fire mission for our mortars while Cortez fired a TOW missile. The explosion illuminated the hillside, a burst of fire and smoke. The mortars followed, pounding the remaining enemy fighters into silence. We both smiled at the explosions, having ended our enemies. We smiled because we were the ones who got to keep breathing and stay alive—at least for one more night. We didn't kill out of hate. We killed out of necessity, like lions killing the hyenas that threaten the pride. While I love my country and feel deeply patriotic, I didn't hear the national anthem in the heat of battle. We simply killed to survive.

The battle was over, and silence crept back over the COP as I sat in my truck. My hands trembled slightly, the adrenaline slowly ebbing away. The darkness remained vast and unknowable, just as it always had. But for now, we had made it through.

* * *

CHAPTER 10

DAY 6, SATURDAY JUNE 25, 2010: HOUSESTEADS TO EAST WALL HOUSES

I awoke early, around 5:00 a.m., but the sunrise had already begun its quiet work on the former Roman frontier. The first light of dawn crept softly over the land, casting a pale glow through the mist that blanketed the hills. I rolled over and found myself face-to-face with my steadfast companion for the night — the wall, just six inches away. I smiled and reached out, my fingers brushing against a moss-covered stone. I was careful to keep my touch light and not wake the ghosts of the legions who had once stood here. The rough surface bore the marks of time, its grooves and notches carved not just by tools but by centuries of wind and rain. In that moment, I felt a connection — to the men who had built it, to the soldiers who had rested here, to the sheer endurance of this ancient monument.

I rolled onto my back, then turned again to gaze north, into the delicate veil of mist. Rolling hills loomed in the distance, softened by the morning haze, and for a fleeting moment, I could almost see them — the Britons moving silently through the landscape, their figures blurring with the mist. And then, a lone Roman patrol. They moved in a single file, heading north into the unknown. The vision brought the ghost of another memory, of the last American patrol I had watched leave FOB Blessing just a month earlier.

I sat up slowly, allowing the cool, damp air to fill my lungs. It was sharp and invigorating, awakening me, not just for the day's march, but for something deeper, something sacred. My breakfast was simple—just water—but it felt more than enough as I immersed myself in the magic of the moment.

The smells of dewy grass and damp earth rose around me, mingling with the faint aroma of the trees that stood watch nearby. I closed my eyes, listening to the gentle whisper of the wind as it danced through the leaves above. I felt truly alive here, and I knew then that this would be one of those sacred moments I would carry with me forever. It was a moment when the past and present all blended.

This was the most alone I had felt in a long time. The nearest farm was at least two miles away, and the visitor center at the fort wouldn't open for several more hours. The silence wrapped around me like a heavy cloak, and I found comfort in the solitude. I paused and allowed myself to breathe deeply, thanking God for preserving me through so much.

We all face close calls in life, fleeting moments when we brush with death and emerge shaken, yet alive. In war, such instances are not uncommon; they happen several times a week, sometimes even a few times a day. I reflected on the countless occasions I had been just inches from death, with vivid memories of explosions and gunfire flashing through my mind.

Perhaps it's this close relationship with death, more than anything else, that leaves its mark on combat veterans. Living in a constant state of hypervigilance, with death always looming nearby, shapes us in ways we can't fully understand until the war is far behind us. We spend too much time like this—dancing with death—and it influences us profoundly, often terribly so.

After a long, hot, dusty patrol, we were finished. Our massive war chariots lined up at the gate, rumbling through the winding path of jersey barriers into the main entrance of Honaker Miracle. There's nothing like the wave of relief that washes over you when you cross the threshold after a patrol. After hours of stress, hypervigilance, and the constant shadow of death, we were finally free — or so we thought. Inside those walls, with barbed wire coiled above and machine guns and mortars ever watchful, it felt like stepping out of a storm into relative calm.

I allowed myself to relax just a little. My mind began to drift, allowing me to take a break from fixation on the highlands we had just come from, my map, and the ever-present calculation of men, vehicles, and spacing. I loved that fleeting feeling of relief, that decompression, even if I knew it could never last.

Today was just another routine patrol — no combat, ambushes, or chaos. We dismounted from our MRAPs, and I began the slow, enjoyable walk to the TOC with my platoon sergeant, SFC Staley. The heat still clung to us, but I took off my Kevlar helmet, letting the air cool my sweat-soaked head. I allowed myself to joke and talk lightly for the first time in hours. It was a rare reprieve.

And then, like a thunderclap in a calm sky, it hit.

I heard a faint, sharp hiss as something tore through the air near my feet. Dust kicked up inches from where I stood. My brain froze, struggling to process what had just happened. For a moment, it felt surreal, like I wasn't entirely there. My eyes locked on the disturbed gravel in disbelief, and then I looked up.

Staley was already several steps ahead of me — in a full, desperate sprint.

I snapped out of it and bolted.

Sharp cracks of machine-gun fire echoed, creating a deadly rhythm that made the air itself feel hostile. More bursts of dust and gravel erupted around me as I ran, tiny explosions marking how close the rounds were landing. My helmet was still tucked under my arm, but I didn't dare slow down to strap it on. My body moved on pure instinct now, adrenaline pumping through every nerve.

This was it. I was running for my life.

I had always wondered what it would feel like to be so close to death that survival became nothing but a frantic, animalistic sprint. And now I knew. My legs burned as I pushed toward the Hesco barriers, the only thing that could shield me from the incoming fire. The distance felt impossibly far—thirty meters might as well have been a mile.

The sound of my boots pounding against the gravel echoed in my ears. Each step felt heavier than the last, and my breath was ragged. My mind was fixated on one thought: don't die.

I dove behind the Hescos, slamming into the hard cover and gasping for air. My heart thundered in my chest, my hands trembling as the first wave of relief clashed with the realization that the battle had only just begun.

The firefight erupted around me, our men returning fire with the same ferocity that had nearly ended me moments earlier.

I pulled out my binoculars, desperately searching for muzzle flashes on the mountain. But I saw nothing. The skirmish had erupted and disappeared like a flash thunderstorm. It was like that sometimes—death and life could coexist in mere seconds.

* * *

A cool morning on the former Roman frontier

I wished I could have stayed there forever—hours, days, maybe even longer. That perch felt like it belonged only to me, as if it had been waiting all this time to offer me this moment of stillness. I didn't want to leave. For a few more minutes, I sat there soaking in the view and allowing the weight of history and the beauty of the landscape to wash over me. I knew, even as I sat, that this would be one of the best moments of my life—a memory I'd carry with me to revisit whenever the world felt too heavy.

With a sigh, I stood, reluctant. But the day was calling, and the road ahead wouldn't wait for me. I began to break camp, rolling up my sleeping mat and securing my gear. As I climbed back over the wall, I couldn't help but imagine whether any of the Northern warriors had ever attempted this same feat, scaling this height and pushing themselves over the edge in a desperate bid to breach Rome's defenses. The thought lingered as I adjusted my pack.

With one last glance back, I turned eastward. Hunger began to gnaw at my stomach, and my thoughts were now focused on the promise of a shop where I could restock my supplies. Yet, even as I walked, the memory of that perch lingered, a quiet sanctuary etched forever in my mind.

I pushed eastward, and the steep terrain continued to challenge my tired legs as I traversed the Roman Knuckles, following the spine of the ancient world, its stones steadfast against the ages. Finally, I climbed to the highest point of Hadrian's Wall, reaching the survey monument near Turret 35a—Sewing Shields.

I paused, letting the cool mist rising from the surrounding land and Broomlee Lough wrap around me like a mysterious veil. The waters shimmered on the north side of the wall, and the vast expanse of green fields and rolling hills stretched endlessly to the north. The view, so achingly beautiful, stirred something within me—a flicker of peace breaking through the storm of my soul. For a moment, I let the weight of the war fall away.

As I continued eastward, the elevation began to soften a bit, with the hills growing less imposing and the spaces between them stretching wider. Yet, the loneliness of the grasslands mirrored my own emptiness deep within my heart. With every step, the wind caressed my face, whispering secrets of the earth and sky, a gentle reminder that I was still alive. The breeze felt light against my scars, yet heavier than any burden I'd carried. I stopped to take a photo—a reminder of this moment—and gazed into my own eyes staring back at me. Dark, unrelenting, and haunted, they seemed to pierce to a depth I could barely fathom. Even now, when I gaze deeply into the depths of my dark brown eyes, it tells me too many things. Looking into the eyes of the man in that photograph unsettles me. He is not the man I once was.

An early morning march.

A sketch looking to the West, near Broomlee Lough.

A thousand-yard stare

Eyes—they are not the eyes of the man I once was. I have seen what men are not meant to see. Death, fear, and the hollowness of life beyond the battlefield I look not at yo , but through you. I have seen Death. I have seen fear. It is now that I know what I am.

I will never again know fear.

I will never again fear Death.

I came so close to death on the battlefields. There, I wouldn't need a bouquet of flowers. Hundreds of rounds of spent brass and the scent of CLP and cordite would have sufficed.

Yet, I live. I breathe now.

I attempt to adapt to the land of the peace.

My ears of war. My scars tell the tale.

I can never forget... And maybe I don't want to...

Such is the truth.

Broomlee Lough and the Roman Knuckles

I was truly alone, with only the sound of my labored breathing keeping me company. It was a sound raw and primal every soldier knows intimately. That breath, shallow and urgent, served as a stark reminder of our nature. It strips us down to the bare truth of what we are: not invincible, not eternal, but mortal, breathing beings. Each inhalation reaffirms our mortality. We are not gods commanding fate; we are fragile and vulnerable men bound to the earth we tread upon.

I needed this solitude more than I cared to admit. In the frontier lands, amidst the ancient silence of the mountains, plains, and deserts, warriors have always come to purge the heaviness in their bodies and souls. These landscapes don't judge or seek explanations — they are vast and enduring, ready to absorb the pain we carry.

It had been over fifteen hours since I'd eaten, and the hollow ache in my stomach was a steady reminder of my body's limits. Hiking under the weight of gear, burning 300 to 400 calories an hour, hunger becomes more than just discomfort—it becomes part of the experience. The gnawing inside echoed my desperation for clarity, for renewal in this harsh, unyielding terrain. This was my penance, my pilgrimage, and I was willing to endure every step of it.

Just as suddenly as the 'Roman knuckles' of the Whin Sill had emerged from the earth, they began to recede, retreating into the landscape as if bowing to the passage of time. Once a proud and unyielding spine of this frontier, these rocky outcroppings faded much like the legions that had once stood atop them into history, swallowed by the perpetual ebb and flow of empires. As the knuckles faded, the wall also began to fall from sight. The wall was my companion for two days, whispering tales of an ancient struggle and acting as my confidant. But now, its presence dwindled, leaving only scattered remnants, forgotten echoes of the long-departed legions.

At Turret 34, Grindon, I paused. The stones of the wall stood there, worn yet enduring, defying time itself. I placed my hand on their cool surface and gazed back at the string of hills—the Roman Knuckles—that had challenged my every step. I thought to myself that this terrain was made for such a line, a line of rock and men bound together by purpose. It was a monument to endurance, a testament to Roman will against the vast, unconquerable wilderness.

As I touched the turret's stones, my thoughts drifted to Afghanistan, to our own defenses in another unforgiving land. While our COPs and FOBs were grounded in the valley floors, we too had, like the Romans, placed outposts on the high ground—small, precarious perches of stone and sweat. I recalled OP Rocky, overlooking the rugged Watapor Valley, its walls built from crude, hastily stacked stones. During the Afghan elections of 2009, I had hunkered down in the blistering August heat, enduring the longest battle of my life—twelve hours of unrelenting chaos. Touching these Roman bricks transported me back to moments of desperation, where even in the age of drones, precision-guided munitions, and advanced weaponry, the humble strength of stone remained a soldier's sanctuary. There was no mortar, no reinforcement—just stone stacked upon stone, a simple yet enduring shield against the war's wrath.

These ancient stones, located on the Roman frontier, reminded me that the act of building walls—whether for protection or defiance—is as old as war itself. Walls of stone and determination. Walls that remain long after the men who erected them have fallen silent. I rested alongside the wall and listened to the wind whispering through its gaps. The stones were not merely remnants—they were memories, layered one upon another, bearing the weight of countless battles and lives, just as I bore mine.

Solitude on the Roman frontier

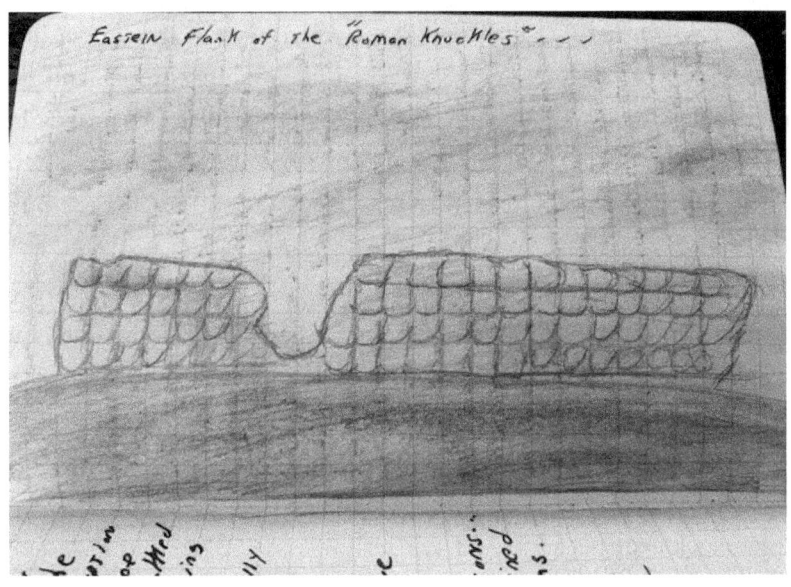

A sketch of one of the last visible fragments of the wall

OP ROCKY, AUGUST 20, 2009

Under the protective veil of predawn twilight, we ascended to our mountain citadel, OP Rocky. The climb was steep and the air thin, but the promise of high ground overlooking the Watapor and Pech Valleys drove us onward. Stones — our ancient allies — shielded us as we took our positions behind makeshift walls, much like Roman legionaries once did on distant frontiers. We watched. We waited.

The valleys below were shrouded in an unsettling silence, a precursor to chaos. The sun crept over the horizon, its golden rays melting the twilight into dawn, bringing with it an oppressive heat that pressed down on us like a second enemy, cooking the rocks and our sweat. The quiet was shattered as battle announced itself with a deafening chorus.

KA-KA-KA! WHIZ-WHIZ! SNAP!

The air turned into a battleground, vibrating with the fury of bullets tearing through it. Tracers streaked like molten threads, and the rocks shielding us sparked and chipped under the relentless onslaught. Some of the enemy fire originated from an elevated position above us. I understood what needed to be done, and artillery was the only solution.

In the distance, the mountains shimmered in the heat, their ridgelines distorted by the rising convection currents. My forward fire team, only twenty meters away, shouted over the gunfire, their voices filled with desperation.

"Sir! We're getting fucked up here!"

Sweat poured into my eyes as I fumbled with my binoculars, trying to lock in an accurate grid. The heat was cooking us on this mountaintop. I grabbed my red Gatorade bottle, wanting so badly to take a sip, but it would have to wait. I placed it on the wall as I stood to get a better view.

ZIP-ZIP-ZIP! SNAP!

Bullets sliced through the air, grazing my ears. I ducked back down, heart pounding. Crimson liquid splattered across my armor, and for a horrifying moment, I thought they had hit me.

"Damn, they got me," I muttered, waiting for the pain to confirm it. My hands frantically searched for a wound, but there was nothing—just the sticky residue of Gatorade. My interpreter handed me the bottle, a bullet hole clean through it. "Oh, fuck, that was for me."

For a few heartbeats, I felt paralyzed. Fear slithered through my veins like venom, and for the first time, I sensed I was a coward. I clutched the dirt beneath me as if it could anchor me to life itself. My mind whispered insidious thoughts: Stay down. Let the others handle it. You don't have to die today.

But something deeper, something primal, stirred within me.

"Fuck it," I growled, the words ripping through the paralysis.

I slid a few meters down, raising my binoculars once more. Through the haze of heat and terror, I caught sight of the telltale flashes of enemy fire—bright and fleeting like fireflies in a twisted nightmare.

"I got you now, fuckers," I whispered. "COP Honaker Miracle, this is Dagger 46. Fire mission request. Grid," I relayed the coordinates.

"Roger, Dagger 46. 155s out of ABAD will respond."

The silence stretched thin as we waited for the artillery. Then, the sky broke open.

"Shot over."

"Shot out."

The earth trembled as the first round crashed down with an ear-splitting CRUMP!

"Fire for effect!" I yelled.

CRUMP! CRUMP! CRUMP!

The mountains roared with the ferocity of unleashed firepower. Each impact shook the earth, sending shockwaves through our bodies. The enemy's fire diminished, then fell silent, buried beneath the wrath of steel rain.

For a brief, fleeting moment, we rejoiced—not in victory, but in survival. We had brought death to those who sought to bring it to us.

* * *

I continued east, and the terrain finally softened as I left behind the unyielding Roman Knuckles into the softening terrain. The jagged hills that had challenged my body and spirit gave way to calm, open ground. My legs welcomed the relief, though my solitude remained unbroken, except for the quiet company of cows and sheep grazing in the lush green pastures. Their gentle movements and low murmurs added a pastoral serenity to the landscape.

Before long, I stumbled upon the ruins of a Roman temple dedicated to the god Mithras. The fragmented stones served as a silent testament to an era when Roman soldiers, far from home, sought meaning in a foreign deity. Mithras, a god whose origins trace back to Persia — modern-day Iran — had journeyed westward with the legions, embedding itself in the hearts of men stationed at the empire's frontier.

Standing there, I felt a strange sense of familiarity. Another prayer from the East, another connection to the lands I had just left behind. The East has a way of captivating and consuming a man whole. Its vastness and mystique leave their mark on all who engage there. Those of us who return are never the same — we carry it with us like a shadow.

I sat among the crumbling stones to rest and reflect. My thoughts drifted to those Roman soldiers, centuries ago, kneeling in this very spot. I wondered what their prayers sounded like as they sought solace or strength from this foreign god. What was it about Mithras that drew them?

I thought of my own time hearing the call to prayer in the East. Those rhythmic and melodic cries, rising over the mountains, carried something both ancient and eternal. Like the Romans before me, I found myself inextricably tied to the East — a bond forged not by choice, but by experience.

JUNE 17, 2009

It was around 1300 hours, and we had been in contact since early morning. Our team sought cover in the heart of the small village of Qatar Kala, nestled deep within the rugged confines of the Watapor Valley. We positioned ourselves near the crumbling stone walls across from the village mosque, its minaret standing like a lone sentinel against the relentless sun. Just a kilometer to the east, our convoy struggled to recover a disabled MRAP on the narrow, precarious road.

At the same time, we held the western half of the valley, ensuring the enemy couldn't reinforce from that direction.

The air was thick and stifling, the oppressive heat pressed down on us like an unseen hand. The sweat pooled beneath my gear, clinging to every inch of my body. For a fleeting moment, I daydreamed, and my eyes drifted to the stream slicing through the center of the village, its waters glinting in the harsh sunlight. It was the only thing in this war-torn landscape that seemed untouched, serene, and timeless.

Then, without warning, the mosque began its Adhan. The melodic call to prayer echoed through the valley, each note a reminder of the world beyond our violence. It rose above the village and into the mountains, carried by the still air, and for a brief moment, it felt as though time had stopped. The sound enveloped us, filling the void left by gunfire and explosions.

Allahu Akbar! Allahu Akbar! Allahu Akbar! Allahu Akbar!
Ashhadu an la ilaha illa Allah. Ashhadu an la ilaha illa Allah.
Ashadu anna Muhammadan Rasool Allah. Ashadu anna Muhammadan Rasool Allah.
Hayya 'ala-s-Salah. Hayya 'ala-s-Salah.
Hayya 'ala-l-Falah. Hayya 'ala-l-Falah.
Allahu Akbar! Allahu Akbar!
La ilaha illa Allah.

God is Great! God is Great! God is Great! God is Great!
I bear witness that there is no god except the One God.
I bear witness that there is no god except the One God.
I bear witness that Muhammad is the messenger of God.
I bear witness that Muhammad is the messenger of God.
Hurry to the prayer. Hurry to the prayer.
Hurry to salvation. Hurry to salvation.
God is Great! God is Great!

There is no god except the One God.

The call to prayer faded, and the valley swallowed its final echoes. Reality settled back in. Suddenly, there was movement. A flurry of figures emerged from the shadowed alleyways of the mud-walled village.

"Holy fuck," I thought, my heart racing. "Is this really happening?"

A large group of women draped in flowing blue burqas appeared, their steps quick yet soundless. Behind them, clusters of children followed, silent as shadows, their tiny feet stirring up dust. They moved like a nervous current, a string of ducklings trailing their mothers, leaving their homes as if propelled by an unseen force. I understood why they were going. The enemy had likely ordered them out, clearing the stage for the deadly theater about to unfold.

It was the silence that struck me the most. Children should chatter, laugh, cry—anything but this muted march. Yet their eyes spoke in ways that words never could. Dark, wide, and unblinking, they revealed everything I needed to know. They left the village so it could be transformed into a battlefield.

"Stay here…" they seemed to say. "Our brothers wait for you in the hills. They moved the moment your armored beasts crawled into our valley. They aim for you now, steady and patient. You will bleed here, and the soil will drink your blood. Stay, and let your fate unfold."

I couldn't see the women's eyes through the latticed screens of their burqas, but their silence carried its own message.

"My husband is up there," they appeared to say. "Stronger than you, unyielding as these mountains. His bullets will pierce your flesh, and his name will resonate in our songs."

The procession passed, leaving a trail of dread in its wake. As I watched them, I felt a sudden and unsettling pang of memory. I was a child again, playing tackle football in the dewy grass of my schoolyard, with the laughter of my friends ringing in my ears. I could almost smell the wet earth, feel the sting of a missed pass, and hear the shrill whistle signaling the end of recess. How far away that life seemed. How blue the sky had been.

And yet, here in this valley, the sky was even bluer. It stretched endlessly, a mocking contradiction to the hell lurking below.

The women and children vanished around a bend, their silent march leaving us behind to face what they knew was coming. Their glances lingered in my mind—haunting and unsettling. They looked at us as if we were already dead. Perhaps they were right. In those moments before a battle, we are neither alive nor dead. We stand on the shores of the River Styx, its dark waters lapping at our boots, its scent filling our lungs. The quiet that followed was deafening, the kind of quiet that belongs only to warriors—a stillness that signals Death's arrival.

The hairs on my neck bristled, and my breath quickened. My men peered through their ACOGs into the highlands, scanning the jagged rocks and flickering shadows. Leaves rustled in the gentle breeze, their movement casting fleeting shapes on the stones. Were they just leaves, or something more? Dread twisted in my stomach like a serpent, tightening with every passing second.

The wind stilled. The world held its breath. Then, the silence shattered.

BOOM! KA-KA-KA-KA-KA! KA-KA-KA-KA-KA! KA-KA-KA-KA-KA!

The dance had begun.

There was not a wallflower in sight.

* * *

Temple to Mithras.

I stepped into the ruins of the ancient temple, and as I walked, I stared at the weathered stones that whispered stories of an empire long gone. I could almost see the Roman soldiers kneeling here at the edge of their world, their prayers rising into the open sky. I wondered what filled their hearts as they prayed here. Likely, their prayers mirrored mine from my own time at war—a simple, desperate plea: to survive, endure, and make it home.

As I wandered among the remnants of the columns, I could almost hear the whispered prayers of warriors, voices lost to the sands of time. Did they pray here daily before marching out into the unknown? Did they kneel at these stones before patrols, offering their fears and hopes to the gods? I thought of my own prayers at the threshold of danger. During war, I found myself more spiritual than ever. Standing at the doorstep of Death, I felt compelled to prepare my soul for whatever lay beyond.

Each day before patrol, I knelt in the quiet of my quarters. The ritual was always the same: strapping on my knee pads, pressing them to the floor, and bowing my head in prayer. I drew the Chi-Rho symbol onto my left hand with the map pen I always had tucked in my armor. This Chi-Rho was the symbol that Constantine claimed to have seen before the Battle of the Milvian Bridge in 312 AD. As the story goes, Jesus had told him, "*in hoc signo vinces*"—"in this sign, you will conquer."

A Chi Rho

I whispered my prayers, low and steady:
"Our Father, who art in heaven…"
A Hail Mary would follow,
"Father, I give my soul to You.
"Help me to be brave.
"Look after my family.
"Give me courage.
"Please, let me protect my men.
"And if I must go, let it be with honor, in a hail of gunfire.
"May my soul rise to the heavens."
I ended with a final invocation: "Saint Michael the Archangel,
defend us in battle…"

In the chaos of combat, I often found myself glancing at the Chi-Rho on my hand. Amidst the deafening gunfire, smoke, and confusion, it served as my anchor—a silent prayer scrawled in ink, asking God to see me through just one more day.

Even today, this temple held the prayers of countless souls. On one of the columns, I noticed coins left by hikers. Maybe they were small offerings to gods or tokens of hope. Some places are forever sacred, etched into the fabric of time by the prayers of those who pass through.

I left the ancient holy ground behind and walked further east as the path unraveled like a green, endless sea sparkling in the sun. The pastures stretched expansive and serene, dotted with sheep and cows grazing without a care, my only companions as I trudged onward. The landscape softened, the earth leveling before me, guiding me toward Black Carts and the remnants of Milecastle 29. The tranquility of the farmland wrapped around me like a cloak, reminding me of the quiet beauty this part of England offers, even to a weary traveler.

Hunger gnawed at me once more, and I found myself yearning for a meal in the village of Chollerford, just a few miles ahead. The march continued, eerily silent except for the sound of my boots on the damp grass. The stillness enveloped me, but after so many miles, it felt unnatural—almost too quiet. I paused to rest at turret 29A, leaning against its ancient stones. For a moment, I closed my eyes and simply listened to the earth.

And then, faintly, the silence fractured. A soft, eerie laughter, distant yet clear, danced on the wind. My skin prickled, and goosebumps rose unbidden on my arms. I stood up, scanning the horizon, my heart pounding. The open farmland stretched endlessly in every direction—empty, utterly devoid of activity beyond the grazing animals. But that laughter... lingered, carried on the breeze, and I could not shake its presence.

There was no one there. No figures hidden in the folds of the hills. No shadow among the ruins. Yet, I knew that I was not alone. The laughter echoed faintly in my mind, whispering secrets of the past. A chill ran through me as I thought perhaps this was the laughter of long-lost legionaries, still holding their posts, welcoming me to their eternal encampment. Then I smiled, the fear melting away like morning mist.

I gazed at the weathered stones of the wall and imagined the weary, dirt-streaked faces of legionaries looking back at me. Their eyes were not those of strangers—they were the eyes of my men, the eyes I saw reflected in the faces of those I had led. And in their gaze, I saw my own. These legionaries, like us, had been sent to the frontiers of their empire. They were tasked with conquering the unconquerable lands and facing peoples who could not be subdued. I felt a kinship with these ancient ghosts. In their silent company, I felt as though I were among my brothers. I was with those who had stood together on the edge of the world.

Near Blackcarts, where I heard the laughter of ghosts

That's the beauty of hiking in places like this: you uncover things that no map or guidebook can reveal. Instead, you find the hidden spots deep within your heart and soul. I realized that this journey offered something unique for everyone who walked it. The Wall had already started unveiling parts of me that I hadn't known existed. The longer I walked, the more I realized I wasn't merely traversing the frontier of a long-fallen empire. Instead, I was entering the uncharted territory of my own warrior soul.

I pressed on my walk, temporarily leaving the fields as I entered the outskirts of the charming village of Chollerford. The path was flanked by stately homes, especially the estate of Chesters Stables. I continued my march downhill toward the North Tyne River and the heart of Chollerford. My heart glowed when I finally came across the diner called Riverside Kitchen. It was my first hot meal in over 24 hours, and I was thankful for the food and rest.

Energized by lunch, I crossed the stunning Chollerford Bridge that spanned the serene North Tyne River. The quiet charm of the village was behind me, and I felt a growing eagerness to leave the paved roads and return to the countryside's sacred embrace. As I walked, a small church appeared — St. Oswald's. Something about its presence called to me, a silent invitation I couldn't ignore. I briefly left the trail and headed for the chapel.

The chapel was modest, perhaps no more than 20 meters long, yet it exuded an ancient, ethereal sense of peace. It was cool; weathered stones seemed to whisper countless prayers and reflections that had flowed through its walls. I was alone here, embraced by the silence, and I took this rare moment to kneel, pray, and contemplate my journey. The rugged miles behind me — the Roman Knuckles, the solitude of the trail — felt like a path meant only for me, a pilgrimage through the frontier of the land and the frontier of my soul.

This site was also rich in its own history, marking the battlefield of an early medieval clash fought in 634 AD. It was here that King Oswald of Northumbria achieved victory over the Kingdom of Gwynedd. The air within St. Oswald's felt heavy with the echoes of past struggles, as if it still carried the spirit of the warriors who had fought here centuries ago.

Inside St. Oswald's

As I crossed the Tyne, a wave of excitement surged through me. I realized that the end of my journey was now within reach—a day or two, perhaps. The anticipation ignited a fire in my chest, driving me to cover as much ground as possible before nightfall. When I arrived at the village of East Wallhouses, I found refuge at the Robin Hood Inn. The locals' warmth and a hearty meal revitalized my weary spirit, and the barkeep kindly allowed me to pitch my tent behind the pub for the night.

Reclining in my tent, beneath the open sky, I felt the earth's pull, as if the ancient Roman frontier was drawing me in deeper. There was no soft mattress here, just the soil, the same ground that had once cradled the feet of forgotten legionaries. They, too, must have rested like this in their time, weary from their own marches. At that moment, I felt a connection, an unspoken kinship, with those soldiers who had walked these lands. I drifted to sleep thinking of my friends Isaac and Seth, eager to reach Newcastle and reconnect, but content to rest for now, one with the land and the echoes of its past.

CHAPTER ELEVEN

DAY 7, JUNE 26, 2010:
EAST WALLHOUSES TO NEWCASTLE

The morning light acted as my alarm clock as the first rays of sunrise kissed the ancient Roman Frontier. I woke up at my campsite just behind the Robin Hood Inn in East Wallhouses. My body felt heavy and sore from the relentless 65 miles I had already marched. However, a new energy surged within me as the prospect of finishing the wall that day sparked a flicker of excitement. My goal seemed tantalizingly close, and I wasted no time. I broke camp, quickly prepared for the day, and set off eastward at a brisk pace.

The trail here was gentle—long, flat, and forgiving. It ran alongside the B6318, aptly named the Military Road. As I walked, the morning sun danced on the waters of Whittle Dene Reservoir, its surface glimmering like liquid gold. I quickened my pace, knowing that each step brought me closer to Newcastle.

Continuing east, I approached the entrance road to Albemarle Barracks. There, I encountered a squad of soldiers on their own forced march. Our eyes met, and we exchanged smiles and waves, a shared camaraderie passing between us. A few of the soldiers noticed the Middle Eastern scarf wrapped around my neck and glanced at my face. I thought they might have recognized me as one of their own. It was a fitting encounter—a meeting of fellow soldiers, their heavy packs and steady pace echoing my own journey. As they passed, I couldn't help but wonder: had any of them served in Iraq or Afghanistan? For a moment, I felt a quiet bond, a silent understanding forged by distant battlefields and the heavy miles traveled.

The scene unfolded before me as I approached the site of what was once the Roman fort of Vindobala, one of the sixteen mighty strongholds that Rome had anchored along this wall. Now, it was just a grassy field on the edge of a quiet farm. Yet, the earth itself seemed to remember its history. The remnants of the fort—piled bricks scattered across the ground—whispered tales of ancient power, of a time when legions stood vigilant on this frontier. I was weary, my body aching from days of relentless hiking, but something about this place called me to stop and rest.

I sat in a forgotten corner of this former bastion of Roman might, leaning against the trunk of a weathered tree. A gentle breeze flowed through the branches above, carrying the soft rustle of leaves—a melody of peace. The grass before me rippled like waves in the ocean under the wind's caress, and I watched it in a tranquil daze. My gaze drifted to the SPQR tattoo on my shoulder, a personal reminder of my connection to the past and to the ideals of Rome that had shaped so much of who I was. The sun shone warmly on my skin, the temperature a perfect 68 degrees, and I was utterly at peace for the first time in what felt like ages.

As I sat there, I reflected on war—not just the ancient conflicts of the Romans and Greeks that had fueled my imagination for so long, but also the battles I had fought in the Pech River Valley, worlds apart yet eerily connected. Here I was, following a path that felt destined, a journey carved from the fabric of time and purpose.

In this moment, surrounded by the quiet strength of Vindobala's memory, I felt the weight of my life's victories and struggles. I was not perfect—none of us are—but my blade of life, though nicked and scratched, still shone in the light. And for that, I allowed myself to celebrate.

Time seemed to pause as I embraced this fleeting yet eternal moment. It seemed like the fort itself was sharing in my silent reflection. In this sacred space, I found what I had been searching for: peace. This ancient fort, Vindobala, gave me yet another important moment to rest my weary body and, more importantly, my soul. This moment would remain with me forever, a treasure etched into my soul.

This moment had me caught between two worlds, alive yet suspended in the memory of how close I had been to death. My eyes landed on a pothole in the road just ten meters away, a perfectly innocent scar on the pavement. It jolted me back to another pothole, one not so innocent, one newly formed when an IED exploded beneath my MRAP.

FEBRUARY 2010

KA-BOOM! An IED doesn't just explode; it consumes. It engulfs one's body, your one's, one's very world. One moment, the ground was solid beneath us, the next, we were hurled into a dark void. The MRAP was lifted clean off the ground, and for a single, eternal instant, I hung there, suspended in a storm cloud of death. I was at the edge of life itself, staring into the abyss. I felt her presence—Death—waiting for me at the threshold. I thought I was going to die in that fraction of a second and, perhaps shamefully, found a strange comfort in the darkness.

I don't remember if my eyes shut; maybe they did for a heartbeat. But I witnessed it all—the world drained of light, emptied of sound. In that brief moment, I wondered if I had already crossed the River Styx, leaving life behind. There was a BOOM—I knew there was, there must have been—but sound, like breath, had been stolen away.

Everything was silent, and I couldn't move. Paralyzed, I thought. Was this what dying felt like?

IEDs possess no loyalty. They don't care if you're brave or cowardly. They consume indiscriminately, devouring bodies, memories, and souls. Unlike soldiers, these weapons feel neither fear nor hesitation. They don't worry about loved ones back home, their brothers in arms, or any future. Their purpose is singular — detonate, maim, kill –. They are the perfect killers, crafted solely for destruction, and they fulfill this purpose with cold, unfeeling precision.

Death plants these devices like sinister Easter eggs, hidden gifts for the unlucky. They sit in the earth, waiting for us to trust the ground beneath our feet. They don't confront us honorably; they don't trade fire or meet us eye to eye. Instead, they deliver their hellish message with a limb-shattering, life-ending roar. These machines of chaos are Death's soulless offspring, waiting to tear apart anyone who steps into their trap.

When the MRAP finally slammed back onto the road, the world seemed to return in pieces. Dust and shards of pavement rained down on us like ash.

At first, I felt nothing. Then came the pain. I didn't scream — panic was the enemy now. Slowly, I assessed myself: my hands moved, my toes wiggled, and — thank God — my genitals were still intact. But blood seeped steadily from my knee, warm and sticky. It wasn't excruciating. I could live with it.

We dismounted, and the scene around me felt unreal, like walking through a half-finished painting. My legs wobbled as I stumbled forward, still not fully grounded in the living world. There was no ambush this time — a small mercy. My eyes caught sight of something in the distance, a faint shape on the horizon. An outpost, perhaps. Salvation.

The medic knelt beside me, his youthful face calm yet concentrated. He leaned down, evaluating the wound on my leg. "You'll be okay, sir," he said with a reassuring nod. "We'll patch you up at that outpost over there. No worries."

"Thanks, man," I replied.

He helped me into another vehicle, and for a moment, I allowed myself to catch my breath. I was alive. Bloodied, shaken, but alive.

* * *

My journey through the lush countryside ended as I descended toward the River Tyne. The downhill path was gentle, yet each step carried the weight of the miles behind me and the anticipation of my journey's conclusion. I entered Heddon-on-the-Wall, skirting the edges of Newcastle, where the Wall itself had long been dismantled, though vanished into the sands of time. Once a mighty testament to the Roman Empire, the Wall had been repurposed in many nearby buildings. As I walked, the spirit of the Wall still lingered, a quiet echo in the land itself.

As I passed through farmlands and pastures, a beautiful black horse standing by the fence caught my attention. She watched me, her dark eyes glimmering in the afternoon light. The horse's gaze was magnetic, drawing me in. I approached slowly, and she allowed me to stroke her powerful head, almost bowing as I did so. Her mane and velvety coat felt warm beneath my hand, and for a moment, it seemed as if time itself had paused. In her presence, I felt a connection to something ancient, something eternal. I wondered if she carried within her the spirit of the Roman war horses that once roamed this land. I patted her strong, muscled body and smiled softly, then bid her farewell, whispering a quiet thank you for the solace she provided me.

Soon, I entered the paved paths of Tyne Riverside Country Park, where the river lay waiting for me. The walk was crisp, and the flat terrain was a welcome gift for my aching body.

The miles seemed to glide by, carrying on my steps' gentle rhythm. Finding a grassy spot, I stopped to rest and laid out my old, trusted

Army poncho. The park was alive with quiet beauty, the river whispering nearby, and the air tinged with the scent of wildflowers.

I stopped to rest and as I adjusted the poncho, my fingers brushed against something tucked into the grass—a note, folded delicately. Intrigued, I opened it carefully and found words scrawled in elegant, feminine handwriting. They read:

I am my beloved's, and my beloved is mine.
(Song of Solomon 6:3 — Old Testament)
Love is patient, love is kind.
It is never jealous or envious.
Love does not demand; love is forever.
(Corinthians 13:4-8)
Not all who wander are lost.
(J.R.R. Tolkien)

The words touched me deeply, their beauty mingling with an unmistakable sadness that seemed to rise from the note. I wondered who this young woman was—was her heart fractured, like mine had been? Tears welled in my eyes, spilling silently down my cheeks. The tears weren't just for her, though. They were for myself too, for the long road I'd traveled, for the losses that could never be reclaimed.

The moment felt surreal, like something lifted from the pages of a storybook—too perfect, too poignant to be real. I refolded the note and slipped it carefully into my Moleskine journal. It remains there to this day, a quiet reminder of that moment when the world seemed to pause, offering me a glimpse of its heartbreak and beauty.

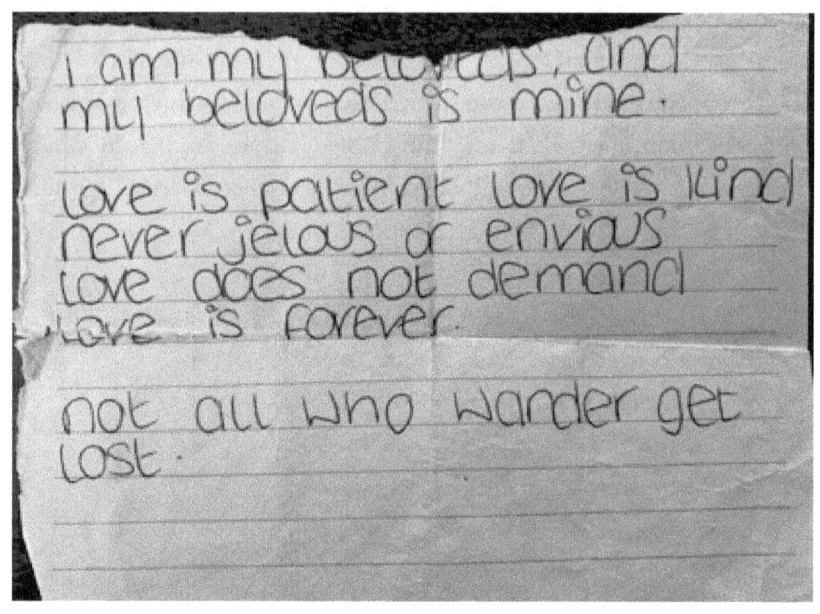

The note

By late afternoon, I entered the city of Newcastle upon Tyne. After nearly a week on the former frontier, it felt strange—almost surreal—to be back in a bustling city. The shift was jarring, like stepping between two worlds. I moved through the city like a tanned, muscular nomad, an outsider amid the urban sprawl of concrete and steel. Gone were my rolling green pastures, the endless sky, and the silent companionship of grazing sheep. Now, I was surrounded by the hum of traffic, flashing lights, and the rhythmic pulse of the civilization I felt only half a part of.

Although I was grateful for the flat terrain, the concrete under my boots gnawed at my already aching feet. I felt increasingly fatigued, but I pressed on, knowing the end of the Wall was only four miles away. However, as the day passed, I decided to rest in Newcastle, opting to tackle the last stretch in the morning.

I found a nice hotel, showered, and took a short nap. The warmth of the bed felt unfamiliar after nights spent on the ground, and for a moment, I wasn't sure if I missed the stars above me or if I was just not used to having a roof over my head. Hunger gnawed at me when I woke, but most of the restaurants were already closed since it was Sunday evening.

I wandered until I found an open pub, settling in with a cold beer in my hand, alone once more. The solitude didn't bother me—I had grown accustomed to it—but the contrast was stark. Days of silence, broken only by the wind and my own thoughts, were now replaced by the murmur of voices, laughter, and clinking glasses.

Then, she sat down next to me.

A beautiful young woman with soft features and kind eyes. She ordered a drink, then turned to me with an easy smile.

"Are you all right?" she asked.

I was caught off guard for a second. I didn't know this was simply the English way of saying hello.

"Umm... yeah," I replied, realizing how weary I must have looked. "I'm just really tired. And hungry. I just hiked Hadrian's Wall."

Her eyebrows lifted slightly. "The whole thing?"

"Yeah."

The pub's kitchen had closed, and I could feel my stomach tightening with hunger. I sighed and asked, "Is there a place nearby to eat?"

She smiled. "Actually, there's a pizza shop close by and I was heading there myself—if you'd like to come along?"

For the first time in days, I let myself relax, allowing warmth to creep back into my voice.

"Yeah," I said, "that sounds great."

Her name was Alexandra, and she had just finished work nearby. We carried our pizza and a bottle of wine to the Gateshead Millennium Bridge, where we sat by the water as the evening breeze rippled across the river. She was curious about my journey, but even more about the war. I found myself speaking about it more openly than I had in a long time. Maybe it was the wine. Perhaps it was the quiet trust between strangers. Or maybe I just needed someone to listen.

As the sun set behind the skyline, casting vibrant colors across the sky, I experienced a rare moment of peace. We finished our meal, exchanged a few more stories, and when the time came, we said our goodbyes.

I walked back to my hotel, feeling full and grateful—not just for the food or the company, but for the kindness of a stranger who unknowingly gave me exactly what I needed—a brief but beautiful reminder that even after war, even after solitude, life still holds moments of quiet connection waiting to be discovered.

My heart glowed as I knew I was close to completing my journey and I fell into a delightful sleep.

CHAPTER TWELVE

DAY 8, MONDAY, JUNE 28, 2010: WALLSEND

I woke just before sunrise, the faint blush of dawn painting the horizon. The air was cool, and I dressed lightly for the first time in days—just a T-shirt, shorts, and running shoes. My rucksack and most of my supplies remained behind in the hotel room. Today, I wanted to travel light for these final four miles to Wallsend. Carrying only my walking stick and camel pack, I chose to run to the end of this long, soul-searching journey.

There's something sacred about an early morning run, when the world is just beginning to stir. My feet struck the pavement softly as I jogged through the deserted streets of Newcastle, the Tyne glistening to my right. The silence felt sacred, broken only by my breath and birds chirping. I passed the towering sign of the "Baltic Flour Mills" building. An English fox darted across my path, pausing for a brief glance before disappearing back into the shadows of an alley. It felt as though nature itself was bidding me farewell.

As I approached my destination, the city yielded to quieter neighborhoods and parks. My pace slowed, and tears began to trace silent paths down my cheeks. I wasn't sure if they were tears of joy, relief, or something more profound—perhaps mourning for the man I was leaving behind, or gratitude for the man I was becoming. Every step felt heavier with meaning as the end of the Wall loomed closer.

Then I saw it: the sign for Segedunum, "the strong fort." My heart soared. I had arrived. Though the fort and museum were still closed, I could see the outlines of the ancient structures through the gates. I stood there, trembling with exhaustion and emotion, gazing at the remnants of what had once been the edge of the Roman Empire. My journey was complete.

I sat on the ground, pulling out my notebook, and penned the words that the Wall had whispered to me over the past days:

Rocks upon Rocks,
Stone upon Stone.
Hands of men, Roman and Britons,
Border of Rome, border of Light.

Further North be dragons,
North lay barbarians.
Here they watched,
There they held.

A wall of stone,
A trench of earth,
Wall of iron,
And Wall of men.
A wall,
A limit.
Perhaps some places can't be conquered.
Perhaps some places should be left dark.

A Legion's whisper,
A drill long forgotten.

The words were a farewell, not just to the Wall, but to the person I had once been. I fell to my knees, overwhelmed by a rush of emotions. I let my tears flow freely, unashamed, and touched the ancient bricks with trembling hands. Then, I leaned forward and kissed the stones — thanking the Wall, grateful it had let me release a piece of the war that had weighed on me for so long. But more than that, I thanked it for helping me discover the new man within.

War had taken so much from me — my innocence and parts of my old self that I would never regain. But this journey along the Wall had given me something in return. It had introduced me to the warrior I had become: a man scarred but not broken, a man who could still feel, reflect, and move forward. This new warrior soul felt like an old friend I was encountering for the first time.

I lingered there, caressing the ancient stones like they were a loved one I would soon have to leave. The thought of parting felt heavy, but I knew this moment couldn't last forever. With a deep sigh, I turned my back on Segedunum and began the four-mile trek back to Newcastle. My muscles ached, and my spirit was both heavy and light. The journey had ended, but the echoes of it would remain with me forever.

When I finally returned to the hotel, I packed my kit, boarded the 11:30 train, and watched the landscape blur past the window. The green pastures and the Wall faded from view. But I knew the Wall had carved itself into my soul. Each stone, each mile, each aching step whispered lessons of endurance, transformation, and acceptance. I realized I was starting a new chapter of my life.

A few hours later, I arrived at Manchester Airport, stepping back into the busy hum of a modern city. The brightness of the terminal and the constant murmur of travelers felt foreign to me. Exhausted, I checked into the Radisson Blu Hotel, located within the airport itself. I felt worlds away from the ancient, wild, untamed frontier I had just left behind. My body ached, and my mind was still tangled in the echoes of my journey. I emailed Seth and Isaac, inviting them to meet me at the hotel.

They arrived later that day, and as I saw their familiar faces, I felt terrible for running away from them. We sat together, with no grand speeches or explanations needed. I apologized for disappearing into my solitude and leaving them behind in the wake of my introspection. But they didn't need my apology. They understood. They had always understood. They knew this wasn't just a long hike or a historical pilgrimage. It was part of the war I was still fighting—the one I had brought home with me, the one I would struggle with for years to come.

That night, we rested and ate well. The next morning, the inevitable parting came. We stood at the terminal, exchanging silent nods, knowing we were saying more than just goodbye. Seth and Isaac boarded their flight back to Michigan, while I turned toward my own gate, bound for Fort Carson, Colorado.

As I walked through the terminal, a strange realization began to settle over me. This journey, this war, this weight—I might never be free of it entirely. But for the first time, I believed that maybe, just maybe, I could learn to carry it.

The wall revealed to me not only my past, but who I was now.

The wall helped me in ways that I will never be able to explain thoroughly in words. I left the wall and hoped that one day I would be reunited with it.

The wall waited.

JUNE 2024: A RETURN TO THE WALL

In June 2024, I returned to the Wall—fourteen years after my first hike. So much had changed in those years. I was no longer the battle-hardened 30-year-old fresh from deployment, nor the man desperately searching for answers along the Roman frontier. Life had reshaped me in ways I could have never foreseen.

I was still in the Army, as I am as of 2025. Shortly after that first pilgrimage along Hadrian's Wall, I deployed again—this time to Iraq. Unlike my previous combat tour, I served as a staff officer at the headquarters of the 4th Infantry Division from October 2010 until October 2011. This experience gave me a different perspective—one from the operational level of warfare. I wasn't directly engaging in battle, but I was observing it unfold, consolidating reports, and coordinating aid from the division level.

When fights erupted along our lines—what we called a TIC (Troops in Contact)—I remained unnervingly calm and laser-focused. It felt like watching a cigarette burn for someone who had quit smoking. The old instincts returned, but in a detached, almost clinical manner. My composure unsettled some of my coworkers.

One day, in the middle of a particularly chaotic TIC, another officer turned to me, his voice edged with disbelief. "Hey man, don't you care? How are you so calm?"

I met his gaze and answered simply, "I do care. But acting frantic won't help the situation." During this deployment, I was able to remain calm much better than most of the soldiers who had benefited from the chaos of close combat. One of the gifts of combat gave me the ability to maintain a calmness, which I called being "The eye of the storm."

By October 2011, I had completed my second—and perhaps last—combat deployment.

From the winter of 2011 to the summer of 2013, I led a headquarters company in the 4th Infantry Division. I loved training with my sole infantry platoon assigned as the Physical Security Detachment. I also loved taking my support soldiers—mechanics, medics, and cooks—through battle drills and training lanes. I embraced the role of being a commander, learning to discover meaning in leadership and striving to share the hard-won lessons I had gained in the valleys, mountains, and deserts.

After my command ended, I left my beloved Fort Carson and the Colorado Rockies in the summer of 2013. I exchanged those familiar peaks for the arid landscapes of Fort Huachuca, Arizona, where I trained as an intelligence officer. Although I missed the thrill of war, I embraced my new role, finding solace in long hikes through the Huachuca Mountains.

In 2014, I completed a short tour at Fort Meade, Maryland, before landing my dream assignment at West Point. Teaching the history of war to those who would one day lead in battle was one of my most fulfilling roles in the Army, next to leading troops in combat.

After West Point, I transitioned into the world of strategic intelligence, serving at places such as the Pentagon. That path ultimately led me to my current role as a Professor of Strategic Intelligence at the National Intelligence University. During this journey, I also had the privilege of pursuing my PhD at Georgetown University.

Amidst all of this, I rediscovered love. I married the brilliant and compassionate Dr. Claire Telford, a native of Antrim, Northern Ireland. Together, we built a family—three remarkable children: Roman, Shannon, and Conan. They became my anchors, my reasons for continuing to strive.

MY RETURN TO THE WALL

I had been away from the war and the Wall for fourteen years. Still, my memories of both lingered, and my heart yearned to return to my favorite place on Earth. So, in the spring of 2024, I made my preparations.

In June 2024, I returned to the Wall with my lifelong friend, Brent Insco, by my side. We first met at thirteen, forging an unbreakable bond through the discipline and camaraderie of karate. Training tirelessly with the North American Alliance of Martial Arts, we earned our black belts together before graduating from high school. Now, Brent was a respected professor and a black belt in Jiu-Jitsu. Although he wasn't a seasoned hiker, his physical conditioning and indomitable spirit made him the perfect companion for this journey.

Brent walked most of the hike with me, but he also understood that there were times when I needed to walk alone to find solitude and reconnect with the man I had been and the man I had become. Throughout this journey, I was reminded of the warm hospitality of those who lived and worked along the wall. One of the best places we stayed was Howgill Cottage Bed & Breakfast. The gracious hosts, Jayne and Karl, were incredibly helpful, welcoming us with comfortable beds, excellent food, and engaging conversation. I will always be grateful to the modern inhabitants of the wall.

Brent and me at Howgill Cottage

Brent and Me at Milecastle 39

The Wall was just as magical as I remembered — perhaps even more so, as if time had only sharpened its beauty, like a long-lost love made clearer in memory. My 44-year-old body was no longer as nimble as it had been in my battle-hardened youth, but it carried me faithfully across the wall and to our various accommodations, walking 100 miles in six days. Every step felt like a reunion with an old friend. The rolling hills, the ancient stones, and the whispers of history along the trail rekindled memories of that first trek. In the stillness, I could almost hear the echoes of the past calling me home.

During my hike this time, the landscapes, terrain features, and — of course — the Wall embraced me like one of the great lost loves of our lives. The Wall was still there, waiting for me as always, and my heart glowed as I laid my hands on its ancient stones.

At Hare Hill, the first visible stretch of the Wall filled me with warmth. The rolling green hills and the little sheep scattered across the landscape rejuvenated both my body and soul, bringing me a joy I hadn't experienced in years. As the trail became steeper, it revealed the expansive horizon of the Roman frontier — the land I had missed so deeply. Brent walked alongside me on that hill and, after a long pause, said, "With all of this beauty... maybe even a warrior can find peace here." He was right.

When the *Roman Knuckles* emerged on the horizon, my heart truly ignited. Those jagged ridges of stone, rising from the clenched fists of the Roman Empire, sent a surge of breath and blood through my body, reminding me of the cost of conquest and the toll on a soldier's soul. They cleansed me — just as they had over a decade ago — washing away my fears and anxieties in a way nothing else could.

Upon reaching the Knuckles, I noticed a small sign nailed to a post. It bore a simple but powerful message.

"Not all who wander are lost."

Something about those words struck me deeply. How had this quote reached me here, on the Wall, after all these years? A shiver ran down my spine, through my very soul, as I whispered it to myself, instantly reminding me of the old note found on the outskirts of Newcastle in 2010. In that moment, the Wall spoke to me. I had wandered since the war, but it assured me that I was not lost.

There is hope for all of us who wander, like ships without wind, after war.

Me on my return to Housesteads Wood

Entering Housesteads Wood was nearly overwhelming. As I stepped under its canopy, I gazed down at my old resting spot on the Roman Knuckles' cliff ledge. It was still there, waiting—just as it had for thousands of years, and just as it had waited for me for over a decade. But this time, a recently cut tree lay where I once sat. I looked down at it and, in my mind's eye, I saw my younger self there—the man I had been, fresh from war, carrying the stench of battle, killing, and death. I stayed there for a while, sitting silently, watching, listening. Listening to the echoes of who I once was.

The Wall continued to bring me beautiful moments and memories from my first odyssey away from war as I trudged eastward. On the second-to-last day of my return journey, I followed the long, flat road of the B6320, just east of Chollerford. I remembered this stretch well and looked forward to reaching my cherished fort at Vindobala once again. A warm sun and a brilliant blue sky welcomed me as I approached this sacred place, and just as I neared the ruins, a rainbow arched gracefully overhead. It felt as if the Wall itself, and the skies above, were embracing me once more.

For a moment, I stood in quiet disbelief, overcome not by the weight of war or the burdens of the past, but by a profound sense of belonging and peace. The serendipity of it all struck me deeply. I paused, took it in, and offered my silent homage to God within the ruins of that ancient Roman outpost.

On the final day of my journey, I woke up in the village of Heddon-on-the-Wall and prepared for my last approach to Newcastle. My older body ached far more than it had at thirty. My right calf cramped violently, forcing me to stop and stretch more often than I would have liked. I worried that a pulled muscle might prevent me from finishing, but with careful pacing and patience, the pain eased, and I pressed on.

My heart glowed as the path gently sloped downward toward the River Tyne. The streets of Newcastle welcomed me back, and as I entered the final ten miles of my journey, a deep sense of contentment settled over me. Passing the Millennium Bridge, I smiled, recalling the meal I had enjoyed there over a decade ago.

A Rainbow on the Way to Vindobala

The sky suddenly darkened, and a hailstorm descended upon me. I rushed for cover under a stand of trees, fumbling for my raincoat. The storm passed within minutes, and the sun broke through again. I cleared the city, following the path along the Tyne and through the outskirts.

As I approached Wallsend, the land offered one last gift. The storm's abrupt end left the blacktop trail steaming, a thin mist rising from the ground like smoke. I smiled as I stepped into it, envisioning the ghosts of my lost Romans, their spirits walking beside me. Tears filled my eyes as I reflected on my first journey along the Wall, my years since, and the life I had lived in between.

I had survived. I had even thrived.

Yet not all of us had.

Some of us drifted upon the ocean of peace, never finding a safe harbor. I thought of Doc Rojas and John Wade, men who left this world far too soon. I prayed for my brothers—that they might live long, peaceful lives and find a solace that war too often steals away.

At the end of the trail, Brent was waiting for me. He had taken the train from downtown Newcastle to meet me at the marker that signified the official end of the path. That marker hadn't been there when I first walked the Wall.

I had returned to the Wall. But in truth, the Wall had never left me.

It had always been with me, its lessons carved into my heart and soul. And as I stood there, having walked its ancient stones once again, I felt something I hadn't experienced in years—complete.

LEAVING WAR

It has been well over a decade since I last heard an enemy bullet tear through the sound barrier around my ears. Since then, life has taken me far from the battlefield. I have fallen in love and married the woman of my dreams. I have been blessed with three beautiful children—Roman, Shannon, and Conan. I have stood in the halls of West Point as an assistant professor, written books, and continued to serve in uniform as an intelligence officer.

Physically, I am thousands of miles and many years removed from war. Yet, every so often, I feel its tremors deep within the shadows of my heart and mind. The memories and experiences of combat still echo, rumbling like distant thunder, reminding me of who I am—who I have always been.

I think I hide it well. Many of us do. We learn to bury the darkness, to keep it tucked away beneath the layers of our daily lives. But in those quiet moments—when the world is still, when solitude presses in—we find ourselves back in the fight.

Ironically, it is in peace that I remember war the most. In those silent hours, the bullets whisper their bittersweet siren's song. In the dim glow of my gym, on a lonely forest trail, beneath a winter sky, or beside a summer campfire—I return. In those moments, I dare to explore the hidden chambers of my soul.

It is there that I reach back—to the war, to the past.

To who I am.

COLD WINTER NIGHT HIKE, GAITHERSBURG MD, 2023

It's below freezing, and fresh snow blankets the ground. Yet, I need to venture out. I seek peace in the darkness. After settling the kids down for the night, I bundle up, grab my rucksack, and step into the chilly night.

The world is quiet—too quiet. The snow absorbs every sound, dampening the streets and muffling even the wind. I am alone now, accompanied only by the rhythmic crunch of my boots pressing into the frozen ground. A full moon hangs overhead, illuminating the night with that eerie glow familiar to all soldiers and hunters—the kind of light that reveals too much and yet not enough.

I tilt my head back, gazing into the dark expanse above me. Orion is there, watching over me, just as he always has. I recall the first time I truly saw him—during a night range at Marine Corps boot camp in 1998 on Parris Island. Since then, he has been a silent guide, a constant presence in my life, always waiting for me in the sky.

I exhale slowly, watching my breath dance in the cold air, illuminated by the moonlight before it vanishes into the darkness. It lingers only for a moment, but it reveals everything I need to know.

I am still here.

I find comfort in the darkness. Here, beneath the moon, I am at peace. The night hides the quiet suburbs around me, blurring reality. And for just a moment, I leave my street behind.

For just a moment, I go back there.

DECEMBER 2009: COP HONAKER MIRACLE

I trooped the line on a cold winter night. The air is crisp and biting, filling my lungs and making me feel alive.

As I walk toward one of our guard towers to check on the men, the silence of the outpost presses in. The night is dead still—only the crunch of gravel beneath my boots keeps me company. I've always loved that sound. Long before the war, even as a kid, I found something satisfying in the rhythmic crunching beneath my feet. I used to imagine a Roman legion marching to the same cadence.

I key my radio to give them a heads-up before I approach. Then, stepping onto the steel stairs, I climb into the concrete guard tower. Inside, SPC Warner and SPC Martinez sit watch, their breath visible in the cold. The air lingers with the sweet, acrid scent of cigarette smoke. They know to smoke away from the gun ports, but on nights like this, nicotine helps keep them awake.

"Hey, sir… good to see you," Warner nods.

I smile and make small talk, then glance out toward the darkness.

"See anything?" I ask.

"Not a goddamn thing," they both reply

I lower my night vision goggles, and the world transforms. That eerie green glow spreads out before me, unveiling a quiet battlefield. Across the Pech River, scattered villages are scattered across the terraces, ascending toward the mountains.

I do not see the earth as men do anymore.

I see it through the eyes of a hunter—piercing the dark, searching. Hunting.

Warner offers me a smoke, and I accept with a grin.

I bring the cigarette to my lips and inhale deeply. The warmth of the tobacco cuts through the cold, offering a small, fleeting pleasure. I don't smoke regularly, but enjoying a cigarette or cigar occasionally reminds me what it feels like to be alive. As I exhale, I watch the thin stream of smoke curl into the air, drifting higher and higher until it vanishes into the night.

I glance at Warner and Martinez, their eyes scanning the terrain beyond the barrel of the 240. They are focused, vigilant. I smile.

I flick away the last embers of my cigarette and turn to leave. "Good luck, boys."

"G'night, sir."

* * *

As of June 2025, it has been fifteen years since I first walked the Wall, and in many ways, that journey has never ended. Each day is another step into understanding the entirety of the warrior's soul. I walk not only to remember, but to rediscover myself. And along the way, I've come to understand what all warriors must eventually face: there is no going back to who we were before our time spent in the fires of god forsaken combat.

Perhaps that is the essence of this story. Maybe it lies at the core of every warrior's journey. Although we can no longer return to who we once were before combat, we can learn to make peace and begin to embrace the individuals we have become. We must continually check in with the constellation of the stars within our hearts and souls to maintain our course, continuing the endless navigation toward a life of peace. However, we must remember that our journey will not always be one of fair winds. More often, we will navigate through the storms of our minds.

I wrote this book because war never fully leaves us. But that does not mean we are lost. There is a path forward for each of us. For me, that path began on the windswept ridges of Hadrian's Wall. The rhythm of my boots beside those ancient stones helped me listen again to my voice deep within. That trail, carved by empire and time, became sacred ground—a place where I could walk beside ghosts and still hear my own heart. This trail is special to all who have walked it—veterans and civilians alike. The Wall continually calls to me, and I will return to walk in the shadows of Rome for as long as I am able.

But peace is not only found on the frontier of the Roman Empire. If you look, you can find it in ordinary life. Your peace can be found within your heart; you must listen. It can be found in the quiet ache of a gym session. In the breathless calm after a hard jiu-jitsu match. In the soft laughter of your children as they roast marshmallows under a summer night's sky, or in the arms of your loved ones. Peace lives in those moments.

The important thing is to search for it. And to know that you are not alone.

I wish you strength and peace. I wish you the very best on your journey.

ACKNOWLEDGEMENTS

Thank you to my dear wife, Dr. Claire Telford, and to my children, Roman, Shannon, and Conan. Your love, understanding, and support give me the motivation to become a better husband, father, and man every day.

To my parents, Claudio and Tereza—thank you for giving me everything I needed to grow into a man and a soldier. To my siblings, Melissa, Claudio Jr., and Rachel—thank you for your unwavering support and for all the letters you sent during my time at war and in peace.

A special thank you to my close friends, Isaac and Seth Williamson, and Brent Insco, who accompanied me on this pilgrimage to the Wall. I will never forget the dozens of miles you walked beside me on the frontiers of the Roman Empire and of my own soul.

Thank you to Mr. Bob Babcock and the entire Deeds Publishing team. You have given a voice to many soldiers like me, allowing us to share our stories of war with honesty and dignity.

To my friend Rachel, thank you for sharing your deep knowledge of history and your sharp editorial eye. Your insight, care, and encouragement were invaluable in shaping this book.

To my fellow warriors under arms, no words can ever fully capture the depth of your sacrifices for our country. You have given more than can ever be repaid or truly expressed in writing.

Lastly, thank you to all who have walked Hadrian's Wall and to those who work to care for it. The Wall will forever live in my heart, and I will continue to walk it for as long as I am able.

ABOUT THE AUTHOR

Antonio Salinas grew up in Allen Park, Michigan. He joined the military in 1998, initially serving in the U.S. Marine Corps before transitioning to a career in the U.S. Army. Throughout his service, Antonio has worked as a topographic intelligence analyst, martial arts instructor, infantry officer, assistant professor in the Department of History at West Point, and strategic intelligence officer, with tours in Afghanistan and Iraq. He continues to serve in uniform and lives near Washington, D.C.

www.ingramcontent.com/pod-product-compliance
Lightning Source LLC
Chambersburg PA
CBHW061150120626
46546CB00005B/1996